What's that ROCK OR MINERAL?

What's that ROCK OR MINERAL?

Tom Jackson

DK

LONDON, NEW YORK, MUNICH, MELBOURNE, AND DELHI

DK LONDON
Senior Editor Peter Frances
Editor Lili Bryant
US Editor Jill Hamilton
Project Art Editor Francis Wong
Pre-production Producer Adam Stoneham
Producer Linda Dare
Jacket Designer Mark Cavanagh
Jacket Design Development Manager Sophia MTT
Managing Editor Angeles Gavira Guerrero
Managing Art Editor Michelle Baxter
Publisher Sarah Larter
Art director Philip Ormerod
Associate Publishing Director Liz Wheeler
Publishing Director Jonathan Metcalf

DK DELHI
Senior Editor Anita Kakar
Editors Susmita Dey, Himani Khatreja
Art Editors Divya P R, Vaibhav Rastogi
DTP Designers Sachin Gupta, Nityanand Kumar
Managing Editor Rohan Sinha
Deputy Managing Art Editor Sudakshina Basu
Production Manager Pankaj Sharma
Pre-production Manager Balwant Singh
Picture Researcher Aditya Katyal

First American Edition, 2014
Published in the United States
by DK Publishing
4th floor, 345 Hudson Street
New York, New York 10014

14 15 16 17 18 10 9 8 7 6 5 4 3 2 1
001–256805–Mar/2014

A catalog record for this book is available from the Library of Congress.
ISBN 978-1-4654-1592-9

Printed and bound by
South China Printing Co. Ltd, China

Discover more at
www.dk.com

ABOUT THE AUTHOR

Tom Jackson is a science writer based in the UK. He began his career as a conservation worker and zoologist, surveying the jungles of Vietnam, capturing buffaloes in Zimbabwe, and working in UK zoos. He has written more than 100 books on science, technology, and nature, and contributed to many more. Tom lives in Bristol, England. His past books include ***DK Eyewitness Books: Science***, ***Animal***, ***Endangered Animal***, ***Spot the Bug***, and ***Help Your Kids with Science*** (with Carol Vordeman).

Contents

Introduction

Rocks and minerals are everywhere. They form the ground we stand on and the solid floor of the oceans, reaching down into the planet for thousands of yards to the Earth's crust. This book will help you identify the most common rocks and give you an insight into the minerals they contain.

The book is arranged in four chapters. The first three deal with rocks and are divided on the basis of the sizes of mineral grains—the crystals you can see when you look closely at a specimen. Each chapter is subdivided by shade—pale, dark, or mixed. The book's simple guidelines will open up the world of rocks for you and let you figure out what you are looking at—a rock layer in a cliff, a pebble on a beach, or perhaps the stone of a building. The final chapter contains the most important minerals. Some are the basis of common rocks, others are valuable ores, and yet more are cherished as gemstones. Rocks and minerals are diverse and can be tricky to pin down as one type or another. This book is a great place to start.

Tom Jackson

What is a Rock?

Rocks—of which about 300 types have been described to date—are defined in terms of how they are formed and their major mineral components. A few rocks are composed of a single mineral, but most contain a combination of different minerals.

Pink granite
The differently colored grains in this granite sample are the minerals quartz, feldspar, and mica.

Rock formation

Rocks are grouped into three families—igneous, sedimentary, and metamorphic—on the basis of the way they form. They occur in a range of sizes, from grains of sand or silt to pieces many feet across.

Basalt sand
Black basalt sand is of igneous origin and, as with other igneous rocks, it forms when magma cools down and solidifies.

Slate pathway
Existing rocks are squeezed and heated by immense forces underground to form metamorphic rocks, such as slate.

Sandstone landscape
Small particles or grains, of solid material build up in layers on Earth's surface to form the sandstone cliffs, such as the rock formation seen here.

What is a Mineral?

Minerals are naturally occurring substances that have a specific chemical composition and are solid in normal conditions. Most of Earth's rocks are composed of just a few minerals—mostly silicates, such as quartz, feldspars, micas, and pyroxenes.

Quartz
The simplest silicate, quartz, is made up of silicon dioxide—a compound of silicon and oxygen.

Orthoclase
The feldspars are a large group of silicate minerals. There are two main types: the alkali feldspars (including orthoclase) and the potassium feldspars.

Biotite
This member of the complex mica group includes potassium, magnesium, and iron in addition to aluminum, silicon, and water.

Augite
This mineral of the pyroxene group contains magnesium, iron, calcium, and aluminum as well as silicon and oxygen.

Identifying Rocks

An important part of identifying rocks is to figure out the conditions in which they were formed. Clues to this can be found in the surface features of a rock.

Texture

In rock identification, the term texture refers to the way minerals combine to make up a rock, and also how the rock's surface feels to the touch. An important aspect of a rock's texture is the size of its grains. Rocks are usually described as being coarse-, medium-, or fine-grained.

COARSE-GRAINED

MEDIUM-GRAINED

FINE-GRAINED

PORPHYRITIC

VOLCANIC

VESICLES

Foliation

When rocks are compressed and subjected to heat, their crystals align into sheets known as foliations. Many metamorphic rocks, including gneiss, are foliated.

GNEISS

Flow-banding

Wavelike bands—evident in igneous rocks, such as rhyolite—reflect the flowing motion of the magma that formed them.

RHYOLITE

Bedding and folding

The deposits that form in sedimentary rocks build up in layers, called beds, that are often visible in rock samples. Frequently, the beds are folded into curves by the forces acting on rocks underground.

Fold formation
Pressure and heat have created folds in these limestone formations on the island of Crete, Greece.

Bed of rocks
Sandstone rocks in various shades of red are arranged in the form of beds at the Zhangye Danxia Landform Geological Park in Gansu Province, China.

Fossils

Rocks containing the remnants or traces of organisms that once existed are known as fossils (pp.100–101). These fossils not only tell us what life on Earth was like in the distant past, but they also help us understand where and when the surrounding rock was formed.

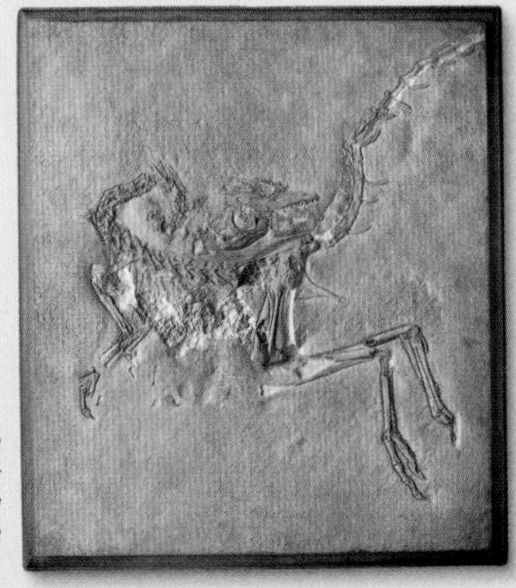

Fossil in limestone
In this fossil of the dinosaur *Compsognathus*, hard body parts, such as bone, have been turned to stone.

Identifying Minerals

Unlike rocks, minerals have a specific set of characteristics. Color, however, is not one of them. While a mineral has a typical color when pure, impurities can create a range of hues. More consistent characteristics, such as shape and luster, can be used to accurately identify minerals.

Mineral shapes

A mineral's shape, or habit, is its outward, large-scale appearance. This is determined by the mineral's molecular structure.

Luster

Not all minerals are sparkling crystals. The way light interacts with a mineral to create its surface appearance is called luster. All lusters can be easily distinguished by the naked eye.

Hardness

While some minerals are very hard, others can be easily scratched or worn. Most mineralogists use a comparative system called the Mohs scale to measure a mineral's hardness. In this system, the numbers 1 to 10 are assigned to ten marker minerals—1 being the softest. A mineral with a higher number will always scratch a mineral with a lower number, so a few simple scratch tests can indicate how hard a mineral is.

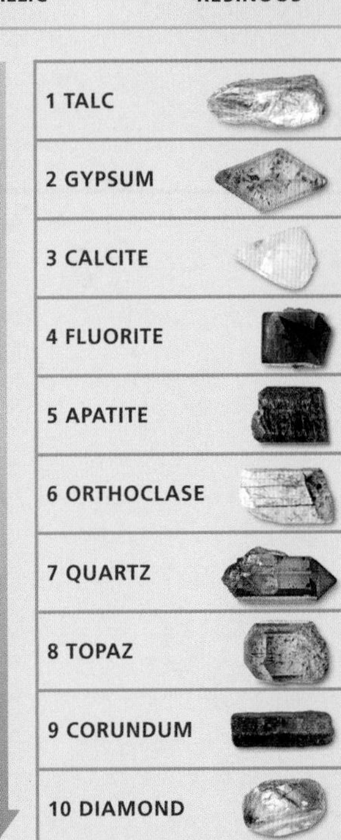

The Mohs scale

Scratch tests based on the Mohs scale can be used to help identify a mineral. For example, a mineral that scratches apatite but is scratched by quartz is rated 6 on the Mohs scale.

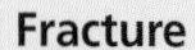

Fracture

The way a mineral breaks, or fractures, is another identifying characteristic. For example, metals often show rough, jagged fractures.

SPLINTERY

EARTHY

Sharp edges, jagged points

HACKLY

Arc-shaped fracture

CONCHOIDAL

Irregular fracture

UNEVEN

Cleavage

Crystalline minerals have specific molecular structures—crystals are built from repeating units with a particular shape, such as cubes or tetrahedra. Cleavage refers to how cleanly and easily a mineral's ultrafine crystal lattice allows it to split. If it splits along a smooth break, the mineral is said to have perfect cleavage. For example, the topaz sample shown here has perfect cleavage. Other minerals have imperfect or difficult cleavage, while some cannot be split at all.

PERFECT CLEAVAGE

CLEAVAGE PLANES

Streak

The color of the powder left behind after a mineral is dragged across a rough tile is known as its streak. While impurities can cause a mineral's color to be variable, the powder produced will always have the same color. For example, this iron oxide hematite has a red streak, while magnetite, another iron oxide, has a black streak. Orpiment has a distinctive golden yellow streak.

ROCK AND MINERAL PROFILES

In this book, the Earth's most common rock types have been divided into three broad categories on the basis of the size of their mineral grains—coarse, medium, and fine. Also featured in this book are some of the most important minerals, which include rock-forming minerals, valuable ores, and gemstones.

1 COARSE-GRAINED ROCKS

The grains and crystals of coarse-grained rocks are easy to see with the naked eye. The grains are at least 1/16 in (2 mm) across and can, on rare occasions, be several inches long.

Pale, Coarse-grained Rocks

GRAPHIC GRANITE

Also known as textured granite, an igneous rock with a higher proportion of feldspars than most granites. Quartz, another component mineral, forms large crystals often with straight sides. This gives the surface a repeating, geometric look.

PINK GRANITE

Pink igneous rock with dark, chunky flecks of biotite mica and grayish patches of quartz. It gets its pink color from alkali feldspars rich in potassium. It is commonly used as a decorative stone.

POLISHED STONE

GRANITIC PEGMATITE

Igneous rock with composition similar to granite, except with larger crystal grains. This gives the rock a rough, chunky appearance. The mica it contains may appear as layered sheets of crystal. Pegmatite sometimes contains large gemstones.

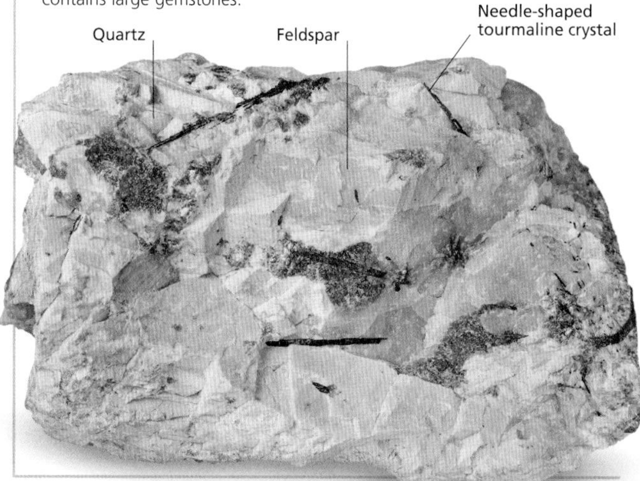

Igneous Rocks

Igneous rocks are formed from the solidification of hot magma, either deep within Earth or when it erupts onto the surface as volcanic lava.

The mineral composition of an igneous rock depends on the chemical makeup of the magma from which it cooled. Rocks composed of quartz- and feldspar-rich magma are generally pale. Those made of silicate minerals containing metals such as iron and magnesium are usually darker.

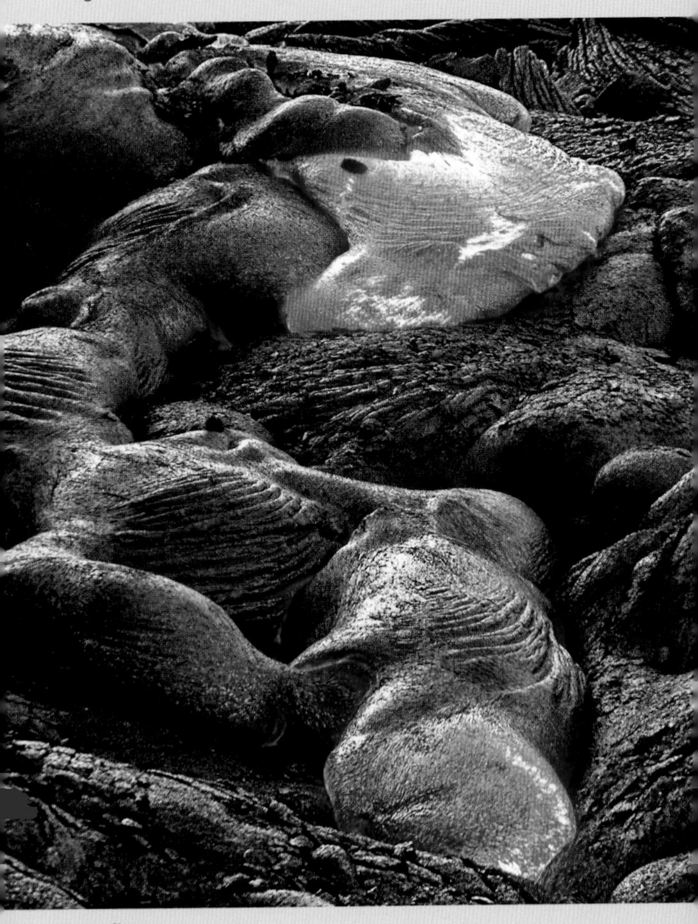

Lava flow

When magma—a hot soup of chemicals, mostly silicon compounds—reaches the surface, it is known as lava. As magma cools down, its compounds form crystals of various minerals.

Intrusive rocks

Igneous rocks that form deep underground are described as intrusive. They cool slowly, which gives minerals time to develop into large, easily seen crystals. As a result, intrusive rocks are usually coarse-grained.

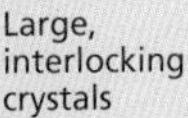

PEGMATITE

Half Dome at Yosemite, US
Made from granodiorite, the Half Dome formed as an underground magma chamber solidified and was then exposed as softer surrounding rocks wore away.

Extrusive rocks

When exposed to air or water, lava cools rapidly and forms extrusive rocks. Unlike intrusive rocks, they typically have tiny crystals, and so appear fine-grained.

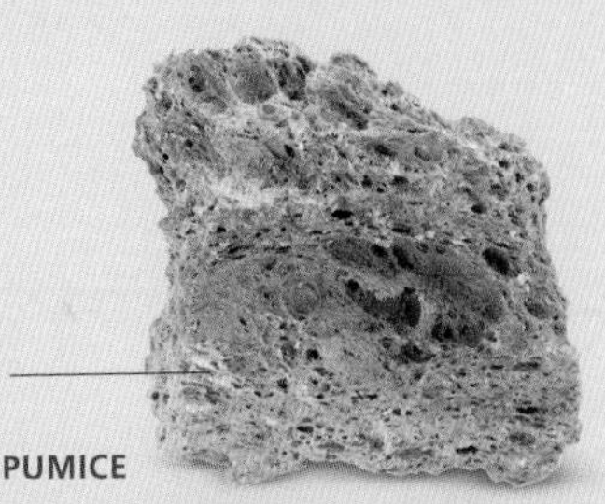

PUMICE

Basalt columns
Hanging basalt columns, like these near Svartifoss waterfall, Iceland, form when a pool of lava cools into rock. As it solidifies and shrinks, the rock cracks into hexagonal columns.

Dark, Coarse-grained Rocks

KIMBERLITE

Igneous rock that is rough to the touch. It is always dark, although there may be a few crystals of calcite. A major source of diamonds, kimberlite is named after the diamond-industry centre of Kimberly, South Africa.

PYROXENITE

Rough igneous rock made almost entirely of pyroxenes. These minerals give pyroxenite a greenish brown tinge. Other iron-rich minerals add to its dark color and also its weight. Pale crystals are rare and small.

POLISHED STONE

SERPENTINITE

Green metamorphic rock formed at shallow depths beneath Earth's surface, mainly from peridotite. Its large, angular crystal grains may line up in bands. The dominant mineral is serpentine. Green olivine, dark hornblende, and reddish garnet may also be present.

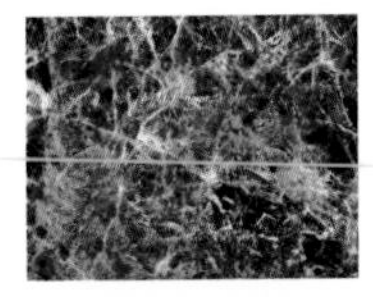

POLISHED STONE

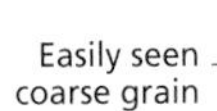

Mottled, patchy texture

ECLOGITE

Rare and unusual metamorphic rock. It has a green-and-red color and visible crystals. The green color comes from a pyroxene called omphacite, while the red is from garnet crystals. The crystals are frequently foliated—lined up in crooked bands.

Red pyrope garnet

Coarse-grained Rocks, Mixed Shades

GRANITE

The most common rock in Earth's continental crust. This igneous rock is typically a mix of dirty pink, black, gray, and white crystals. Mica gives smooth, weathered specimens a waxy sheen, while rougher rocks get a sparkle from the quartz.

TOURMALINE PEGMATITE

Igneous rock similar in color to granite (above). Compared with granite, it has much chunkier crystals of orthoclase—a type of pinkish alkali feldpsar—and quartz mixed with elongated, prismatic crystals of a much darker tourmaline. The tourmaline may have a striped appearance.

AMPHIBOLITE

Metamorphic rock largely made of amphibole minerals, most commonly the dark, green-black hornblende. The dark grains are normally aligned with each other. The occasional pale color is due to calcite and feldspar crystals, with a flash of red from garnet.

Plagioclase

Coarse texture

Garnet crystal

Amphibole crystal

DIORITE

Igneous rock often seen as a black and white stone. It is made up of two-thirds alkali feldspars that form creamy crystals. Dark minerals such as hornblende and biotite make up the other third. It is sometimes called black granite. Unlike granite, however, it contains no quartz.

»

GRANODIORITE

Cross between granite (p.24) and diorite (p.25). Although it has the same rough texture and speckled appearance as granite, this igneous rock is a little darker because of its mix of blue-gray plagioclase feldspars and pink alkali feldspars. It can be pink or white.

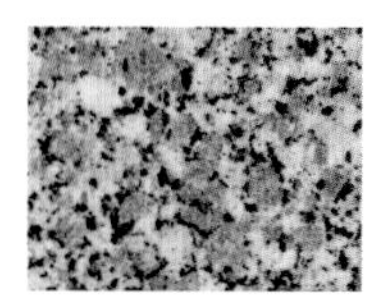

POLISHED STONE

CONGLOMERATE

Distinctive sedimentary rock of varying colors, with smooth stones set in a cementlike matrix. The stones can be anything from pebbles to boulders, but an individual specimen generally contains stones of similar size.

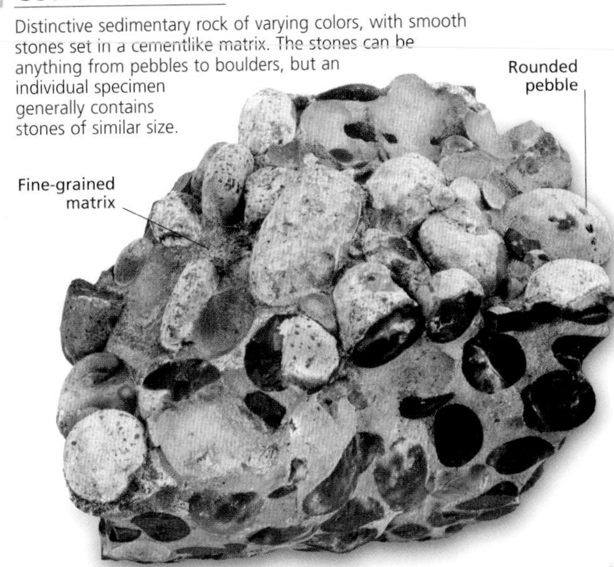

BRECCIA

Sedimentary rock similar to conglomerate (above) but made up of angular, fragmented stones instead of smooth ones. These are held in a fine-grained matrix. The fragments come in a wide range of sizes and colors.

2 MEDIUM-GRAINED ROCKS

Medium-grained rocks have grains from 1⁄256 in (0.1 mm) to 1⁄16 in (2 mm) in size—visible with the naked eye or under a simple magnifying lens. These rocks may contain a small number of larger crystals called phenocrysts.

Pale, Medium-grained Rocks

QUARTZITE

Commonly grayish yellow but can also be pink, orange, or red in color. Quartzite is a metamorphosed sandstone, in which loosely connected sand has melted to form a denser, heavier, and harder rock made of tightly fused quartz crystals.

ANORTHOSITE

Gray igneous rock that forms deep in Earth's crust. It is not common at the surface, and when it does appear it is usually as an enormous mass. Anorthosite is mostly made from pale, calcium-rich plagioclase, which occasionally appears as large, distinct crystals.

TUFF

Soft igneous rock formed from ash thrown out by a volcano. Tuff is generally uniform in color, ranging from pale yellows to browns and grays. It frequently has layers, formed where larger, heavier ash particles fall to the ground before the lighter grains.

»

MARBLE

The most familiar metamorphic rock. Its chief minerals are calcite and dolomite, giving the rock its typical white color, but as a result of impurities, marble comes in just about any color and pattern. Marble can be cut and polished easily, making it a common building stone.

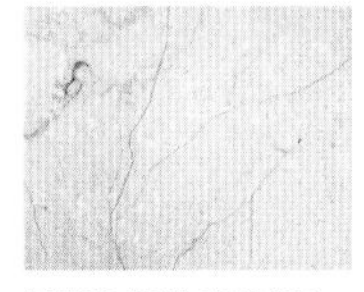
BUILDING STONE

KYANITE SCHIST

Yellowy gray metamorphic rock. It gets its name from the blue mineral kyanite, which can be seen as small blade-shaped flecks among twisted layers of quartz and feldspar.

GREENSAND

Soft sedimentary rock with a dirty green tinge. This sandstone gets its color from glauconite—a mineral formed from organic debris chemically altering rock debris in shallow seas. Greensand has a soft surface and well-sorted grains, with similar-sized grains cemented together.

ARKOSE

Generally pink sandstone with high feldspar content. This sedimentary rock is made up of angular grains. In older specimens, the pink color, coming from iron oxide impurities, darkens to gray.

SANDSTONE

One of the most common sedimentary rocks seen on Earth's surface. Sandstone—formed from grains of sand cemented together—often houses fossils. It can show ripples formed when the sands were blown by wind or washed up by water. Sandstone varies in color from cream to yellow and to orangey red.

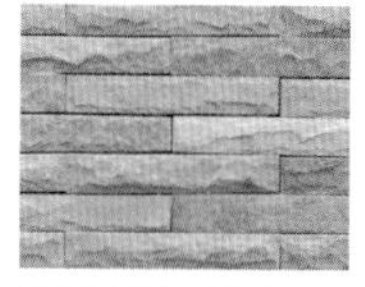
BUILDING STONE

RED SANDSTONE

Red sedimentary rock that has rounded sand grains when formed in the desert and more angular grains when formed in shallow seas. In both cases, drying cracks, ripples, and cross-bedding—where the angle of deposition has changed—can be seen.

BUILDING STONE

GRITSTONE

Sedimentary rock formed from large and spiky sand grains, originally from ancient riverbeds. Together, the grains form a very hard type of sandstone, composed mostly of quartz. Different impurities make gritstones yellow, brown, pink, or gray.

BUILDING STONE

OOLITIC LIMESTONE

Sedimentary rock that is pale yellow in its pure form. It consists of spherical or egg-shaped calcite structures known as ooliths. The calcite is deposited from mineral-rich water.

BUILDING STONE

Sedimentary Rocks

About 80 percent of the rocks found on Earth's surface are sedimentary. These are formed by the accumulation in layers of minerals and other rocks.

Over millions of years, sediments are compressed by the weight of new deposits, which cements them together to form solid rock. Sedimentary rocks are softer than igneous ones because they are held together with a chemical cement instead of heat-fused crystals.

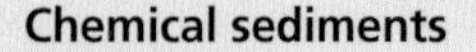

Chemical sediments

Some rocks, such as rock salt, form from chemicals dissolved in water. As this water evaporates, the chemicals precipitate, sink to the bottom, and form a layer of sediment.

ROCK SALT

Clastic sediments

Most sedimentary rocks, including puddingstone, are formed from clasts—crystal grains and fragments broken from larger rocks. Clasts range in size from boulders to silts.

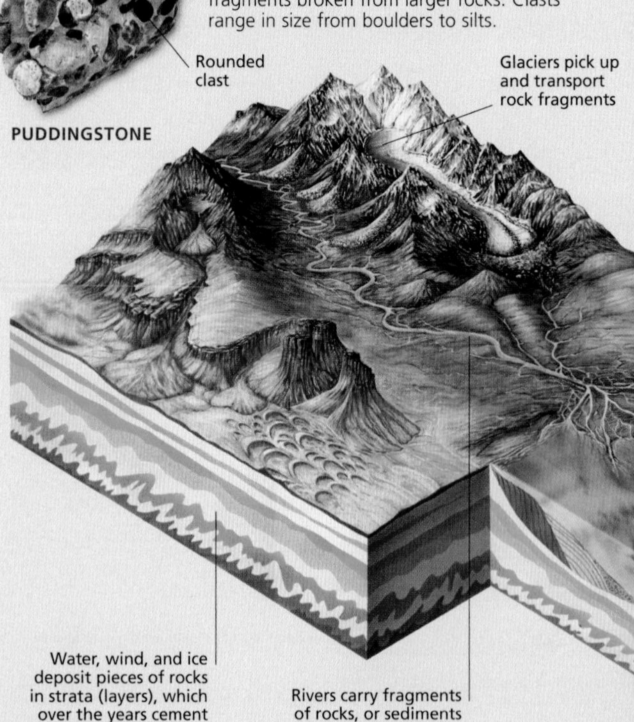

Weathering

All rocks gradually weather, or wear away. Water trickles through rocks, reducing some minerals to chemical sediments. Water, ice, wind, and gravity also breaks rocks into smaller pieces—known as clasts when they become part of other rocks.

Weathered wonder

The Grand Canyon in Arizona, is a product of erosion and weathering. The Colorado River cut through 1 mile (1.6 km) of solid sedimentary rock to create this rock formation.

Erosion

Pieces of rocks can be removed by wind, by moving ice (such as glaciers) or by running water, especially rivers. The process by which rock fragments are loosened and carried away is called erosion.

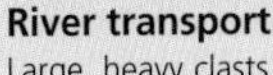

River transport

Large, heavy clasts are not carried far by river water. However, lighter silt particles move great distances with the current, before settling to the bottom of a river mouth or seabed.

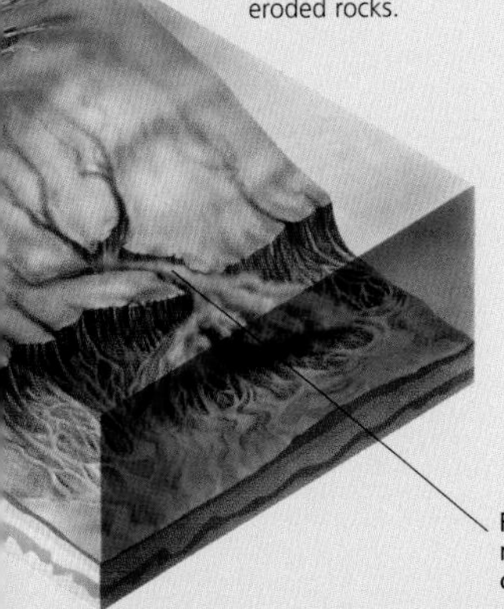

Formation of sedimentary rocks

These rocks—found on or near Earth's surface—are formed in layers from pieces of weathered and eroded rocks.

Diagenesis

Sediments form rocks through the process of diagenesis. In this process, deposited grains are compressed and any water squeezed out, leaving behind chemicals that cement the grains to form a rock.

Dark, Medium-grained Rocks

GABBRO

Igneous rock with greenish tinge from augite and other pyroxene minerals. Gabbro has a highly variable mineral content, but it lacks quartz. Some large crystal grains of pale minerals, such as calcium plagioclase, may be seen among the dark pyroxene minerals.

POLISHED STONE

PERIDOTITE

Olivine-rich, heavy igneous rock that is usually dark green. The dark color is due to the presence of pyroxene minerals. Specimens that are especially rich in olivine (and have fewer pyroxenes) will be a paler green.

DOLERITE

Dark gray igneous rock with composition similar to that of gabbro (opposite), but the minerals occur in different proportions. The dominant mineral is calcium-rich plagioclase, which occurs as inclusions—tiny crystals. These are surrounded by pyroxenes that give the rock its dark color.

LAMPROPHYRE

Dark brown rock, sometimes black. It has a rough matrix of small grains, sometimes studded with larger crystals of pale orthoclase. The matrix of this igneous rock is rich in metals, especially magnesium, iron, and potassium minerals.

MUSCOVITE SCHIST

Metamorphic rock with a large amount of a type of mica called muscovite, which appears as folded layers of fragile white crystal. This rock shares the same general mineral composition and structure as other schists.

FOLDED SCHIST

Metamorphic rock squeezed into ripples and folds by great pressure and heat. These folds may crumble as sheets of silvery mica break away from the rock. The rock specimen itself may also split easily.

ORTHOQUARTZITE

A form of sandstone made from small, rounded grains of quartz sand cemented together. The color of this sedimentary rock depends on other minerals present with the quartz. It is generally gray, but it can also be pink, white, or yellow.

IRONSTONE

Sandstone or limestone that has a very high iron content and typically formed in the distant past. This sedimentary rock is often formed from ooliths—rounded structures coated in iron minerals—that make it red or brown.

Medium-grained Rocks, Mixed Shades

LAYERED GABBRO

Type of gabbro (p.38) in which dark, metal-rich minerals form as layers between paler plagioclase feldspars. The main metal in this igneous rock is iron, but the bands can also include chromium, titanium, and platinum.

LEUCOGABBRO

Black and cream igneous rock. Sometimes coarse-grained, it is a mix of plagioclases and pyroxenes, and very little olivine. This composition makes it less dense and lighter than other gabbros.

CARBONATITE

Consists of carbonate minerals, which is unusual for an igneous rock. Specimens are generally a creamy yellow but have wavy bands of greens, browns, and oranges. The grain size is highly variable.

PORPHYRY

General name for a range of igneous rocks with varying compositions. All porphyries have a medium or fine-grained background matrix filled with conspicuous larger crystals of other minerals, often quartz or feldspars.

POLISHED STONE

»

MIGMATITE

Mixture of rock types, typically metamorphic schist or gneiss contains bands of hard, quartz-rich igneous rock, such as granite. The quartz and feldspar from the granite appear as pale streaks and veins in the dark, mica-rich host rock.

GRANULITE

Metamorphic rock named for its uniform grain size. It is similar to gneiss in composition, but unlike gneiss it is not banded. Granulite is generally dark, with paler flecks of plagioclase feldspars. Garnets may add a pinkish hue.

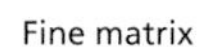

SKARN

Metamorphic rock formed from limestones and other sedimentary rocks made from calcite. The calcite ranges from white to pink to brown and is split by darker silicate minerals, especially by jagged veins of greenish dolomite.

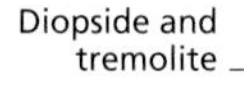

SYENITE

Mix of pink, black, and creamy gray. This igneous rock may be confused with granite (p.24), because it is found in the same upland areas. However, syenite generally has smaller grains and lacks clear, twinkling quartz crystals.

BANDED GNEISS

Common metamorphic rock characterized by a broken banding of light and dark minerals. The crystal grains are flattened, or foliated, as a result of the immense pressures that created the rock.

FOLDED GNEISS

Similar to banded gneiss (opposite), except with more distinct bands twisted into curved and rippled layers. Sometimes, a large distinct mass of pale feldspar crystal known as an augen or eye is seen.

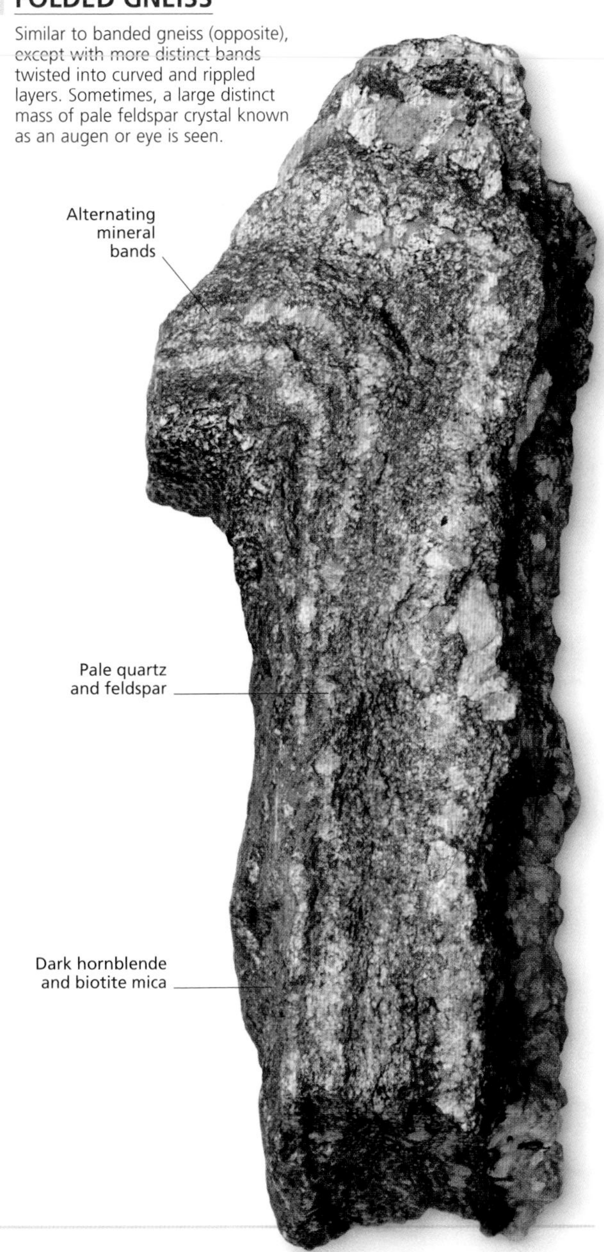

3 FINE-GRAINED ROCKS

Most fine-grained rocks have microscopic grains, which are smaller than 1⁄256 in (0.1 mm). In some cases, these rocks do not have grains at all. Instead, they form a seamless aggregate of crystals, or are amorphous or noncrystalline, like glass.

Pale, Fine-grained Rocks

SCORIA

Igneous rock often found cracked and weathered into rubble. This hard, prickly volcanic rock is reddish when new but darkens with age. No crystals can be seen, and it has hollow, rounded cavities, or vesicles, where gas was trapped as the rock cooled.

PUMICE

Volcanic rock that looks like a stone sponge. Like scoria (above), pumice is an igneous rock filled with gas vesicles. However, it contains silica-rich minerals, which make it quite soft and light. Some pumice specimens can float in water.

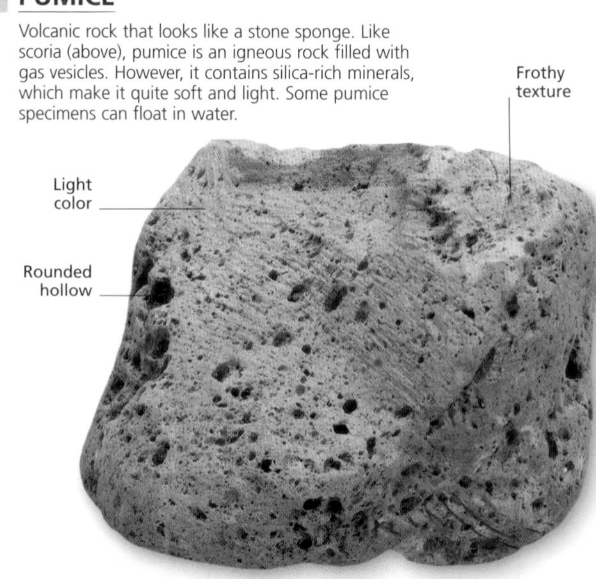

TRACHYTE

Igneous rock made chiefly of alkali feldspars, which form a pale gray matrix. The matrix can also be yellow or pink. It has dark phenocrysts (large crystals) of metal-rich minerals.

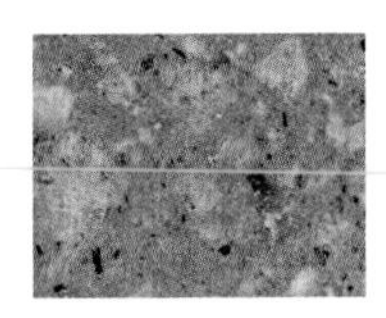

POLISHED STONE

SOAPSTONE

Soft, greasy metamorphic rock made largely of talc. Most soapstones are white, green, or yellow-brown, but they can be darker. This flaky rock is often built up of thin sheets, which can be scratched away using a fingernail.

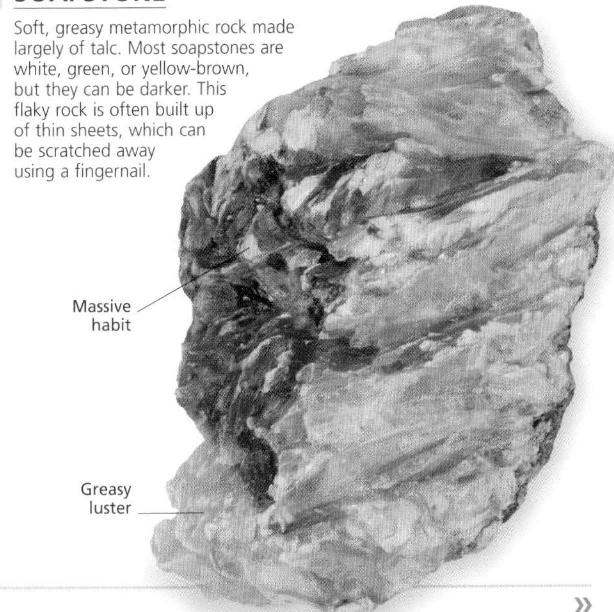

LIMESTONE

Sedimentary rock, usually dirty white but can be pink or gray. It can be made from microcrystalline calcite—too small to differentiate into grains—or shells and other remains of living things. It is very often a mixture of both.

BUILDING STONE

TUFA

Sedimentary rock often stained red by the presence of iron oxides. Tufa is made up of minerals laid down by water from hot springs or desert lakes. The minerals precipitate as the water evaporates, forming gnarled tubes and towers.

DOLOMITE

Gray, yellow, pink, or pale brown sedimentary rock. It forms from limestones in which the calcite has been replaced with dolomite, a carbonate mineral containing magnesium as well as calcium. The mineral is in a microcrystalline form that makes the rock very hard and gives it a uniform texture.

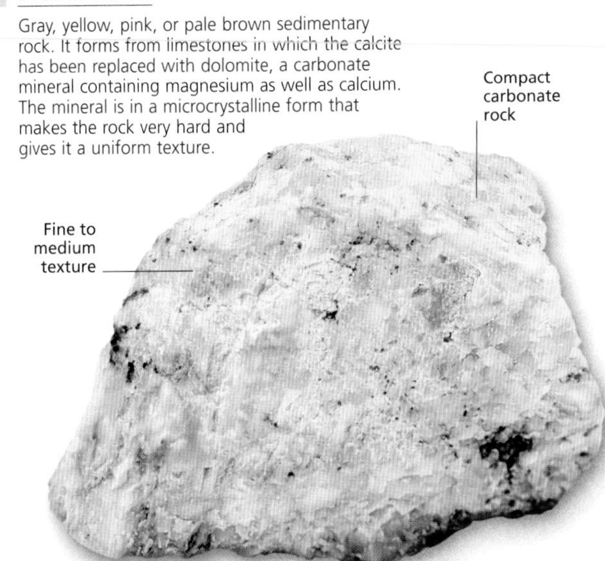

CHALK

Soft white sedimentary rock made almost exclusively from shells of microscopic sea creatures. Chalk has a powdery feel and is often damp to the touch due to the presence of clay minerals. Chalk leaves a layer of powder when rubbed on harder surfaces, although the "chalk" used for writing is generally gypsum.

GREEN MARL

Smooth, hard sedimentary rock, composed of clay and carbonate minerals. Its grains are only visible under a microscope. The green color is derived from organic matter mixed into the rock-forming sediments.

ROCK SALT

Sedimentary rock form of common salt, or sodium chloride. There is generally a reddish tinge to the white crystals, due to the presence of iron oxides, but pink and even blue forms exist as well. Rock salt specimens are very delicate, not least because they dissolve in water.

ROCK GYPSUM

Sedimentary rock that may appear as curving fibers or plates, and also as lumps of tiny crystals. Composed almost entirely of gypsum, this rock is white, yellow, or pale brown. It has other minerals—such as salt and iron oxides—or different rocks running through it.

»

DIATOMITE

Soft, dirty white sedimentary rock. Rough to the touch, it often has large pore spaces—enough to allow some specimens to float. It is made of the silica-rich remains of microscopic algae called diatoms. Diatomite can be crushed easily. Swedish scientist Alfred Nobel used this powder as the stabilizing element in dynamite.

LOESS

Smooth, crumbly, sedimentary stone rich in clay. Loess is pale in color and may appear as a tightly packed earth rather than a rock. It is found in deep, dry beds, and were originally formed in deserts.

Dark, Fine-grained Rocks

BASALT

Dark, compact igneous rock that is generally rough and dull, but some glassy specimens are smooth and shiny. Its crystals are so tiny that they cannot be seen using most microscopes. The most common rock on Earth's surface, basalt forms from rapidly cooled lava.

BUILDING STONE

VESICULAR BASALT

Essentially the same as basalt (above) but filled with vesicles, or cavities, made by gas bubbles trapped inside the rock as it solidified. The presence of these spaces means that, when weathered, this igneous rock has a very rough surface.

ANDESITE

Dark volcanic rock, composed mainly of plagioclase feldspars. An igneous rock, andesite contains the same minerals as diorite (p.25), but it cools more rapidly so its crystals are smaller and only visible with a lens. Some specimens have a fine matrix of dark andesine with larger crystals of paler oligoclase.

OBSIDIAN

A striking igneous rock composed of naturally occurring volcanic glass. It is often jet black due to iron impurities, although red and brown varieties exist as well. The lustrous glass cracks to form edges that can be sharper than surgical scalpels.

RHYOLITE

Gray-brown microcrystalline igneous rock. It looks similar to flint (p.68) because some of its minerals form natural glass. Rhyolite may have bands that show the flow of the magma from which it originated; other forms are pale gray, even pink, with many dark phenocrysts.

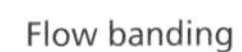

VOLCANIC BOMB

Lump of rock formed in midair from lava ejected by a volcano. These igneous rocks fall from the sky and cool in flight into an elongated shape. Brown or red, weathering to a yellow-brown color, they can be 2½ in (6.5 cm) to 3 ft (1 m) long.

HORNFELS

Dark metamorphic rock that forms when a range of parent rocks is exposed to great heat. Hornfels is a hard rock with a uniform surface, and its mineral composition is varied. It can appear glassy or matte, and it fractures into flakes, leaving rounded ridges on the surface.

BLUESCHIST

Metamorphic rock with a bluish tinge to its dark gray, often banded, surface. The blue comes from the mineral glaucophane. A similar rock with more epidote and chlorite is known as greenschist. Blueschist is found in mountainous regions, where it forms from the heat of magma acting against older rocks.

SLATE

Lead-gray, rarely black, metamorphic rock known for the way it splits into flat sheets. It forms from shales (p.65) and mudstones (p.66). Slate has few surface features because its tiny crystals are aligned into sheets.

BUILDING STONE

PHYLLITE

Metamorphic rock with a waxy sheen due to high mica content and a green tinge because of chlorite. Although similar to slate (above) in the way it can be split, phyllite has larger crystals, so it breaks into thicker, rougher sheets.

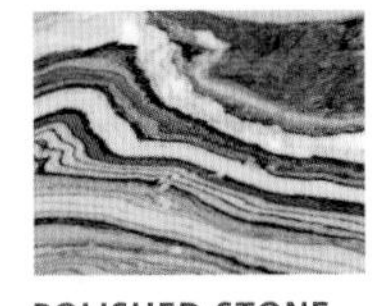

POLISHED STONE

Metamorphic Rocks

The mineral composition of rocks can be changed without re-melting when they are subjected to extreme temperatures or pressures, or both. Such altered rocks are known as metamorphic rocks.

Any rock—whether igneous, sedimentary, or metamorphic—can undergo metamorphosis. Usually, however, the rock is so transformed by the process that it is hard to identify what it originally was.

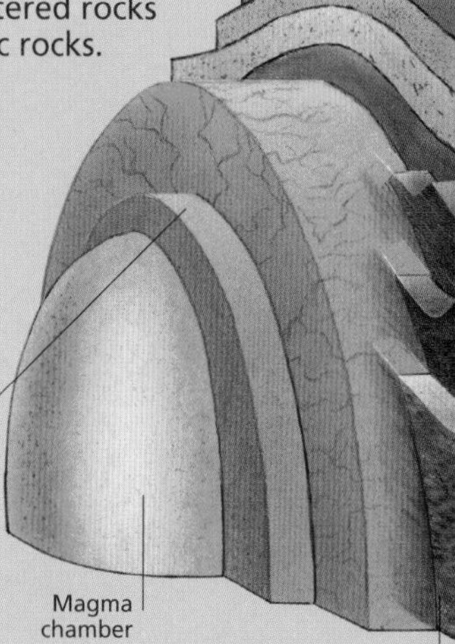

When magma intrudes into rock, it creates high temperatures and low pressures. The surrounding rock is altered in a process called contact metamorphism

Heat and pressure deep in Earth's crust can effect widespread changes in the rocks there. This is called regional metamorphism

Types of metamorphism

The different types of metamorphism include contact, dynamic, regional, and shock. In shock metamorphism, the great heat and pressure generated by a meteorite impact results in the formation of new types of rocks.

Metamorphic grades

The term metamorphic grade describes the intensity of metamorphism. It can refer to the amount of increase in temperature or pressure, or both. The three broad metamorphic grades are low, medium, and high.

Metamorphosis of shale
This sequence shows the various types of metamorphic rock into which the sedimentary rock shale can be transformed by the increasing application of heat (from left to right) and pressure (from top to bottom).

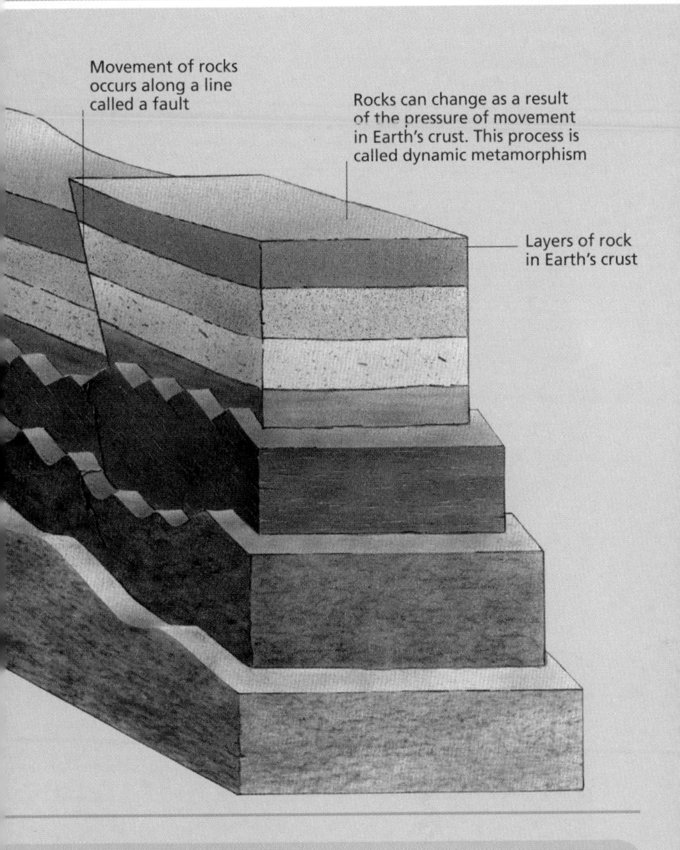

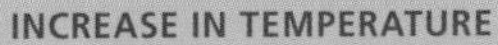

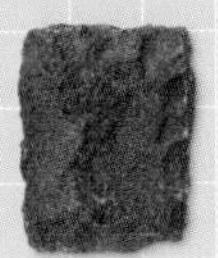
HORNFELS

SCHIST

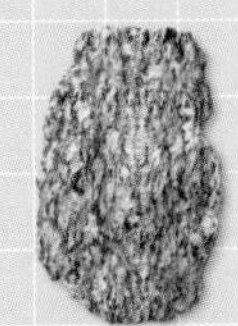
BANDED GNEISS

MIGMATITE

RED MARL

Hard, red-tinged sedimentary rock that breaks into chunks instead of layers. A close relative of green marl (p.54), its uniform red color is due to the presence of iron oxide in the mud from which it formed.

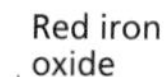

SHALE

The most common sedimentary rock, made of tiny clay grains. Shale is generally gray—darker varieties indicate the presence of organic matter mixed with the original clay. Shale splits easily into sheets along bedding planes. It often contains fossils.

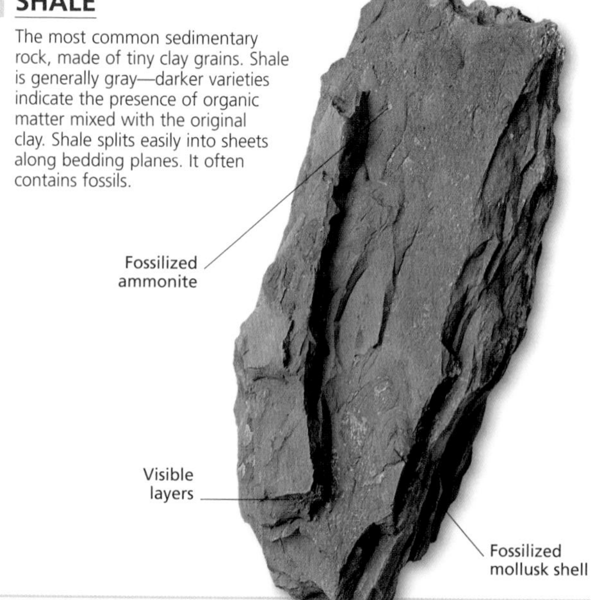

OIL SHALE

Oily, dark brown or black sedimentary rock that produces a strong petroleum odor when freshly broken. The color and sheen are due to kerogen, a tarlike substance derived from plant and animal matter in the rock-forming clay. The presence of kerogen makes it a commercially important oil-producing rock.

»

MUDSTONE

Sedimentary rock made from minute grains of waterborne clay mixed with the remains of algae and other life forms; the latter make the rock dark in color. Unlike siltstone (opposite), this gray or black rock can break into smaller flakes and layering is rare.

SILTSTONE

Sedimentary rock, made from silt with a grain size bigger than that of clay but smaller than sand. Siltstones often show bedding planes and even ripples, where successive layers of silt have been deposited. Most siltstones are dark due to organic matter in the sediment. Fossils are common.

GRAYWACKE

Hard sedimentary rock, mostly fine-grained but with a mixture of larger sand particles also present. It is gray in color, but brown and black varieties are also found. The rock is also called dirty sandstone.

TILLITE

Sedimentary rock with dark gray clay matrix filled with a wide range of other rock fragments. It is formed from the debris dumped by glaciers. Also known as boulder clay, tillite forms from whatever rocks a glacier erodes as it descends a mountainside.

FLINT

Nodular form of chert (opposite) that occurs in areas of marl, chalk (p.53), and limestone (p.52)—all softer rocks that erode to expose flint. This gray sedimentary rock fractures to form sharp rounded edges, which can be used as cutting tools.

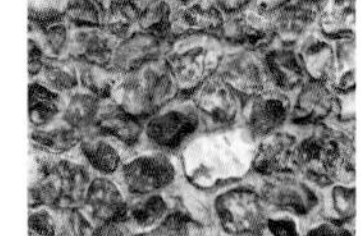

BUILDING STONE

CHERT

Sedimentary rock made entirely of microcrystalline quartz, or chalcedony. Chert has a smooth, glassy surface, especially when fractured. Generally gray or brown, chert can form as layers but is often found as nodules—a nodule is a rounded lump surrounded by another rock.

BOG IRON

Sedimentary rock formed in swampy areas from spring water rich in iron minerals. It consists of misshapen nodules of iron ores, such as red hematite and rusty brown goethite, typically mixed with mudstone (p.66).

COAL

Dark brown or black sedimentary rock, mined for fuel. Coal is a carbon-rich rock formed from the remains of woody plants. It is generally quite soft and a little greasy to the touch due to hydrocarbon tars and oils. Hard coal, such as anthracite, has a glassy surface.

TEKTITE

Once described as meteorites, these rounded, glassy nodules are now thought to be formed by the heat of a meteor strike melting surrounding rocks. Tektites are rare and generally less than a couple of inches across. They are rich in silica and other oxides.

Fine-grained Rocks, Mixed Shades

AMYGDALOIDAL BASALT

Striking form of the common igneous rock basalt (p.57) in which gas vesicles become filled with large crystals of white minerals, especially quartz and zeolites. These minerals get into the rock in water trickling through cracks. The water can also turn dark iron minerals in the rock rusty red.

LATERITE

A bed of loose pink, yellow, and red earths, as well as hardened nodules. Formed when iron and aluminum oxides mix with sand grains, laterites are found in warm, damp climates. They are soft enough to cut into bricks or crush into gravel.

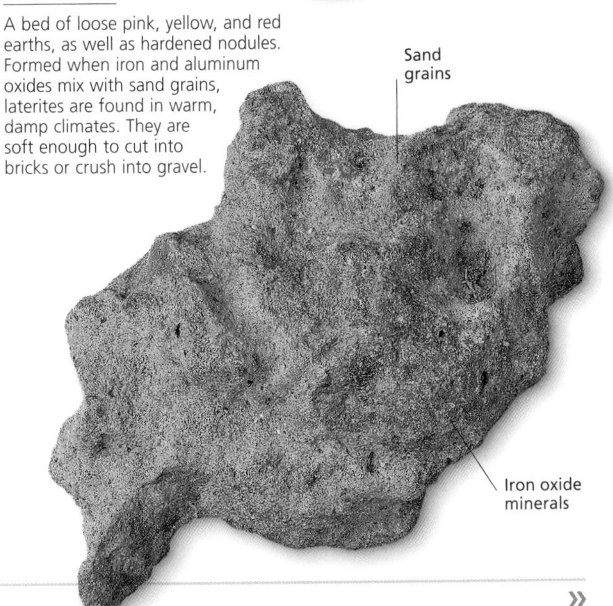

TRAVERTINE

Porous, white sedimentary rock. It is formed by calcite precipitating out of spring water, usually in caves or around waterfalls. Although similar to tufa (p.52) in composition, travertine forms in layers and often comprises rounded structures cemented together.

POLISHED STONE

DACITE

Igneous rock formed near volcanoes, containing nearly equal proportions of quartz and plagioclase feldspars. Dacite is generally dark gray but may contain patches of pink where certain plagioclase feldspars have solidified more slowly than the surrounding fine-grained matrix.

BANDED IRONSTONE

Sedimentary rock made from alternating layers of iron minerals and gray shales (p.65) or cherts (p.69). The iron layer may be rich in a red oxide (hematite), a black oxide (magnetite), or a brown mixture of the two. Banded ironstones are ancient rocks believed to have formed in the oceans about 3.5 billion years ago, when oxygen was first released into the waters.

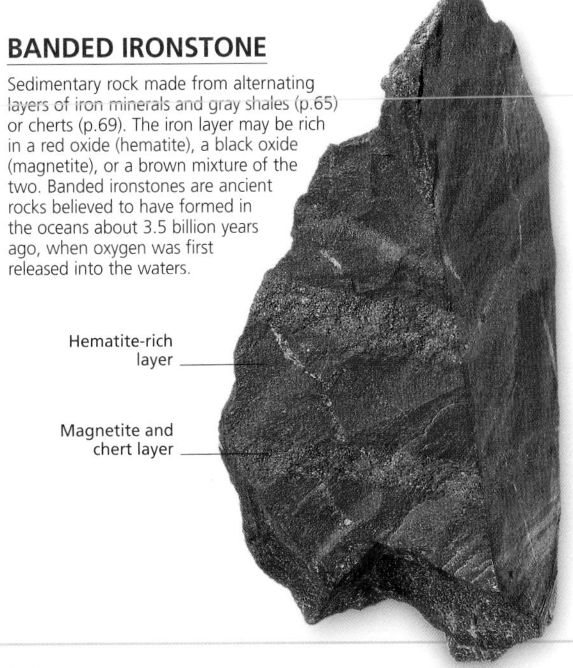

MYLONITE

Metamorphic rock with no specific mineral composition, but characterized by rippled and twisted bands of light and dark running through it. These bands are made by the flattening and stretching of mineral grains.

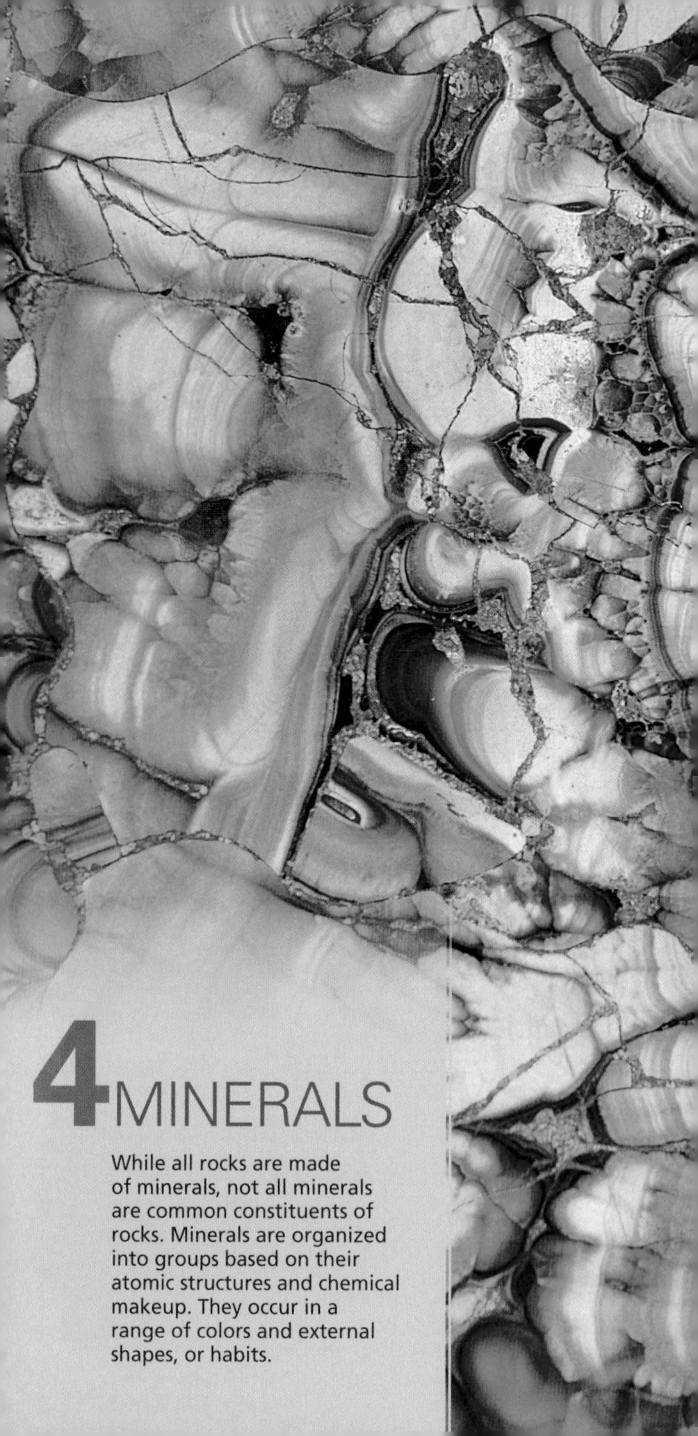

4 MINERALS

While all rocks are made of minerals, not all minerals are common constituents of rocks. Minerals are organized into groups based on their atomic structures and chemical makeup. They occur in a range of colors and external shapes, or habits.

Silicates

Most of Earth's crustal surface is made up of silicates. The minerals in this group contain varying combinations of oxygen and silicon, along with other substances.

QUARTZ

One of the most common rock-forming minerals. The glassy, prism-shaped crystals are colorless when pure. Milky white specimens are more common. Impurities produce dozens of other colors. In rocks, quartz generally occurs as visible white or gray crystals and sand grains.

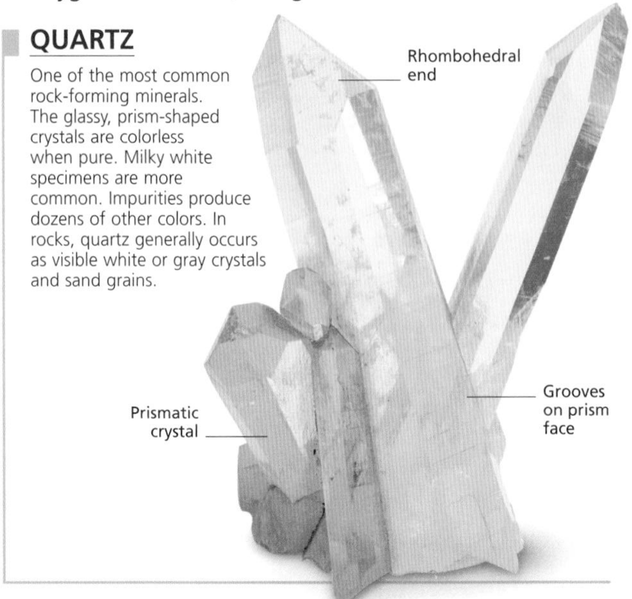

AMETHYST

Pink or purple gemstone variety of quartz (above). Large crystals may have a pyramid-shaped peak. The striking colors are due to iron replacing some of the silicon inside the crystals.

SMOKY QUARTZ

Dark brown or black variety of quartz (opposite), also known as cairngorm. The color comes from defects in the crystal structure produced by natural radiation. Like all varieties of quartz that produce large crystals, smoky quartz forms prisms with pointed tips.

Milky quartz

Smoky quartz

ROSE QUARTZ

Ranging from a cool pink to rosy red, this collectible form of quartz contains trace amounts of titanium, manganese, and iron. Rose quartz is often translucent, due to the presence of tiny bubble spaces in the crystal lattice.

CHALCEDONY

A variety of quartz in which individual microscopic crystals combine to form rounded nodules, fibers, or tubes. Chalcedony is white when pure, but impurities give it a wealth of alternative hues; some are highly prized. Unlike shiny crystalline quartz (opposite), chalcedony has a waxy appearance.

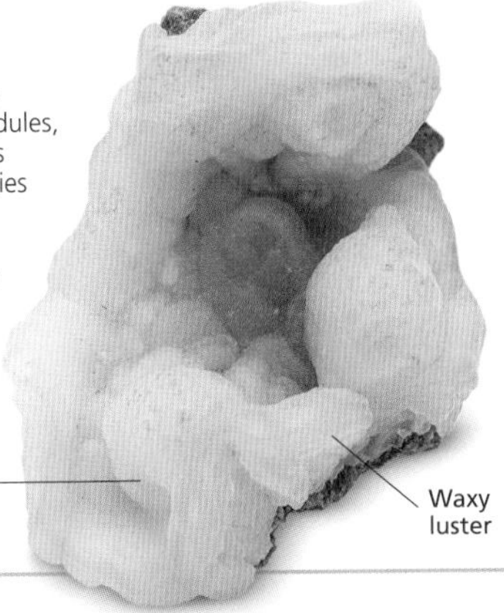

Gems

A gem is any mineral or rock that is used in jewelry or as decoration. Flawless and symmetrical gemstones are rare and, therefore, highly prized and valuable.

There are dozens of gems available, with most minerals having at least one gem variety that is attractive enough to be used in jewelry. Most gems are brightly colored, tough, and durable. They shine when polished and sparkle when cut.

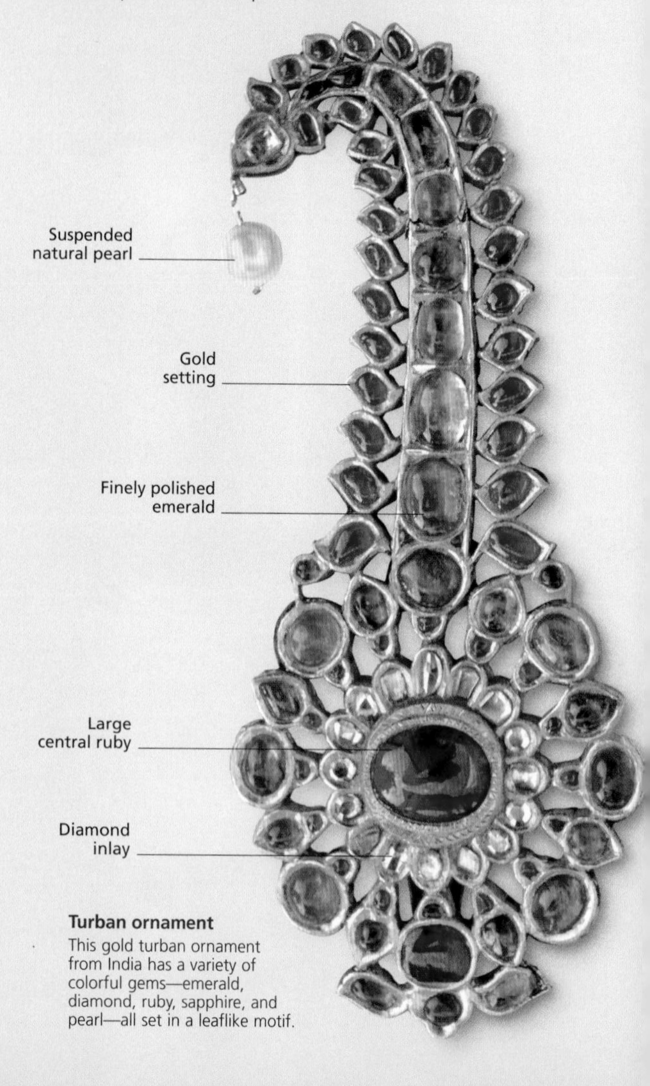

Turban ornament
This gold turban ornament from India has a variety of colorful gems—emerald, diamond, ruby, sapphire, and pearl—all set in a leaflike motif.

Precious stones

The most popular gems are the four so-called precious stones: ruby, diamond, emerald, and sapphire. Pearls are also classified as precious stones, although they are formed by shellfish and are not strictly minerals.

DIAMOND

RUBY

EMERALD

SAPPHIRE

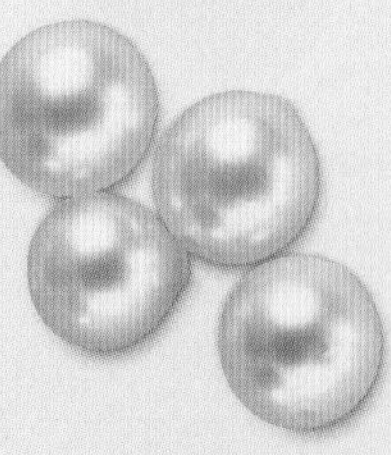

PEARL

Gem cutting

Gems are highly prized for the way they sparkle. This is due to the optical property of refraction, whereby light bends when passing from one medium to another. A crystal can be cut in such a way that the light always reemerges from the upper surface. This gives the gem a dazzling sparkle.

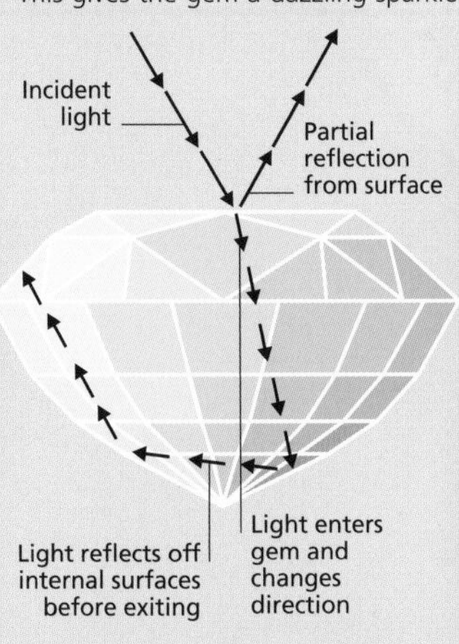

Internal reflection

When light shines into a highly refractive crystal, such as a diamond, it will reflect internally several times before coming back out.

Diamond cutting

The hardest mineral in the world, diamond is difficult to cut—only a diamond can cut another diamond. It must be cut carefully because a wrong incision could be a costly mistake.

AGATE

Semiprecious form of chalcedony, forming in nodules and characterized by concentric rings when cut in cross section. The colors of agate's rings vary depending on impurities but are often golden, brown, white, blue, or red.

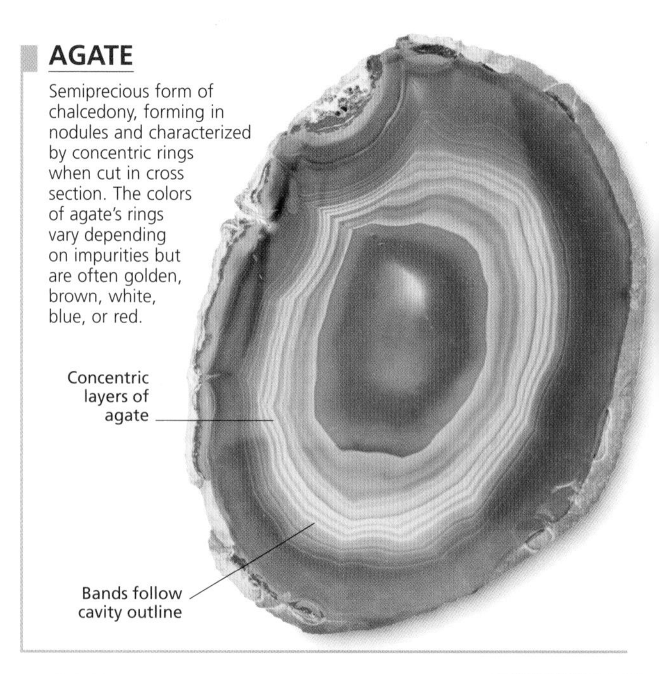

OPAL

Semiprecious mineral that occurs as smooth, glassy stone rather than as crystals. Pure opal is colorless and transparent. However, most stones contain impurities and appear opaque with a shimmering, pearly coloring, mixing blues, yellows, and reds.

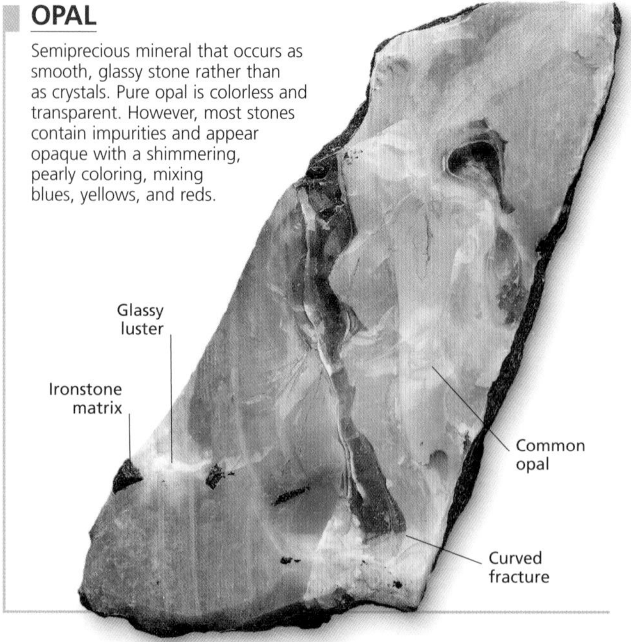

ORTHOCLASE

One of the most common rock-forming minerals, orthoclase belongs to the alkali feldspar group. It has a subtle pink color—as seen in granite (p.24) and other igneous rocks. Orthoclase's glassy crystals often form prismatic grains, but they can also appear as sheets or in larger blocks.

MICROCLINE

An alkali feldspar that can be yellow, brown, pink, or red. It often forms prismatic or tabular crystals. Microcline has a glassy luster, and it is hard enough to be used in jewelry, especially the green variety known as amazonite.

ANORTHOCLASE

An alkali feldspar with glassy prismatic or tabular crystals that can be colorless, white, gray, pink, yellow, or green. Anorthoclase is often seen as twinned crystals. In rocks, it usually appears as fine grains.

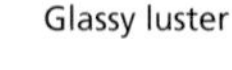

ALBITE

Feldspar mineral intermediate in composition between the alkali and plagioclase feldspars, appearing as tabular or platy crystals, often twinned. Albite gets its name from the Latin for white, but some specimens can be colorless, pink, or green.

OLIGOCLASE

The most common of the plagioclase feldspars. It occurs in granular form in rocks, while larger specimens are generally glassy and massive. Often pale gray, oligoclase can also be red, green, brown, or yellow in color.

ANORTHITE

Common plagioclase feldspar, occurring as small grains in iron-rich rocks. It appears as brittle prismatic crystals of rich pink, red, blue, white, or gray. It is very common in Moon rocks.

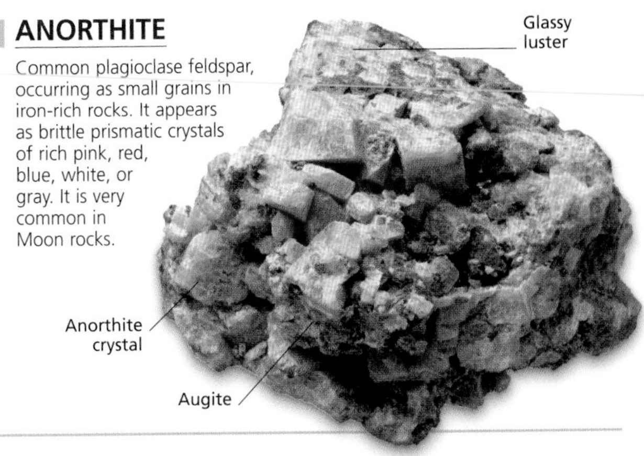

LABRADORITE

Plagioclase feldspar that exhibits an iridescent surface. Labradorite is mainly blue, but its surface reflects a shimmering rainbow of colors. The mineral is seldom found as single crystals. Instead, it occurs as a compact mass of microscopic crystals.

ANDESINE

Plagioclase feldspar named after the Andes mountains, where it is found in andesite lavas. Andesine is glassy gray-white when pure but can also be a darker yellow-brown. It occurs in massive form or as flat tabular crystals.

»

SERPENTINE

Group of distinctive green minerals, which can also have yellow or gray hues. Serpentine has a greasy or waxy luster. Pure specimens are generally massive or form as plates or columns.

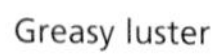

CHRYSOTILE

Yellow-white fibrous mineral, and the main constituent of asbestos. Chrysotile is very soft and crumbles in the fingers. Its thin fibers, which form in the veins between rocks, give the mineral a greasy luster.

TALC

The softest mineral of all. It is the main ingredient in the dry astringent talcum powder used to soothe inflamed skin. Talc can be white, green, gray, or brown. It can occur in a crumbly massive form, or as fragile fibers or flattened sheets.

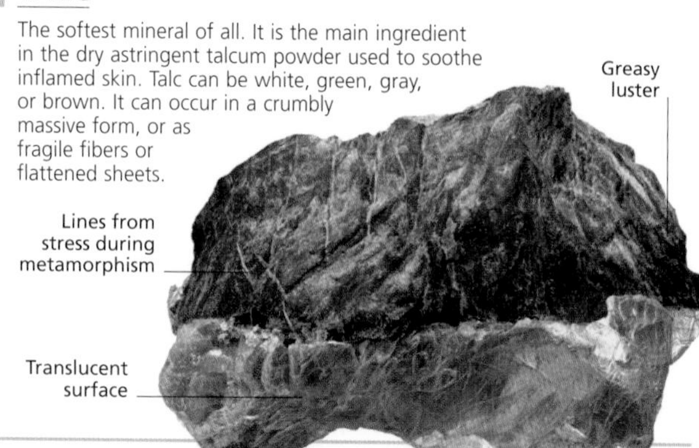

MICA

The name given to a group of minerals, the most abundant member of which is muscovite or common mica. Micas typically occur as flat, platy specimens that flake easily. Micas are usually colorless—large sheets were traditionally used as window panes—but can also be brown, gray, green, or red.

BIOTITE

Dark mica mineral found as large, flat, tabular specimens or stout prismatic crystals. Biotite is black when rich in iron impurities but fades to brown and yellow when magnesium dominates. Specimens are fragile and split into flat sheets.

»

BENTONITE

Sandy gray mineral, with a uniform color and dull surface. One of the most common clays, bentonite is the basic material for earthenware and pottery. It forms soft, earthy masses with crystals that are too small to see.

KAOLINITE

Clay mineral used to add sheen to paper, furniture polishes, and paint. Kaolinite occurs as pale gray, earthy masses that are powdery to the touch or in harder massive forms.

Powdery texture

Earthy luster

Granite matrix

DIOPSIDE

Variable pyroxene mineral generally occurring as a green prismatic crystal. It can also be colorless, blue, or brown. Crystalline specimens are often square in cross section. Diopside can also occur in columnar form, as thin sheets, and as massive, glassy aggregates.

AUGITE

The most common pyroxene, almost always seen as a dark mineral ranging from green to black. Augite specimens are opaque with a rather dull luster. Large crystals are prismatic and square or octagonal in cross section.

Dark, nearly opaque crystal

Volcanic tuff matrix

HORNBLENDE

The name given to a small group of dark minerals rich in metal impurities. Hornblendes generally have a green hue but can also be brown or black. Large crystals have a glassy luster and grow in a bladed habit—elongated with a flat top.

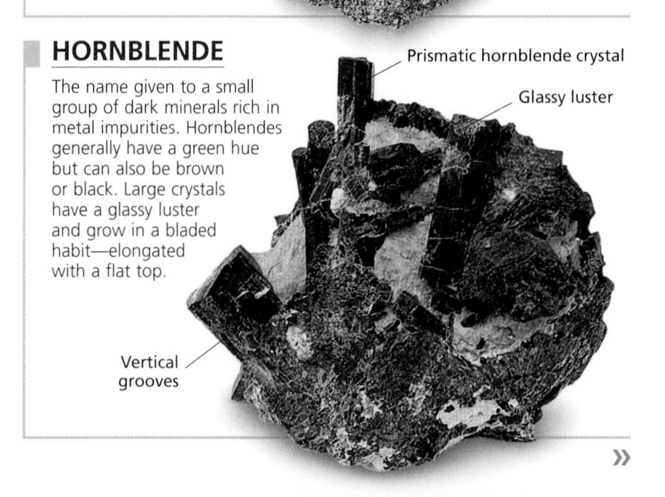

»

TOURMALINE

A group of 11 minerals with blue, red, green, and colorless varieties. The most common tourmaline is the dark, opaque prismatic crystal found in pegmatites. Its crystalline form is highly prized.

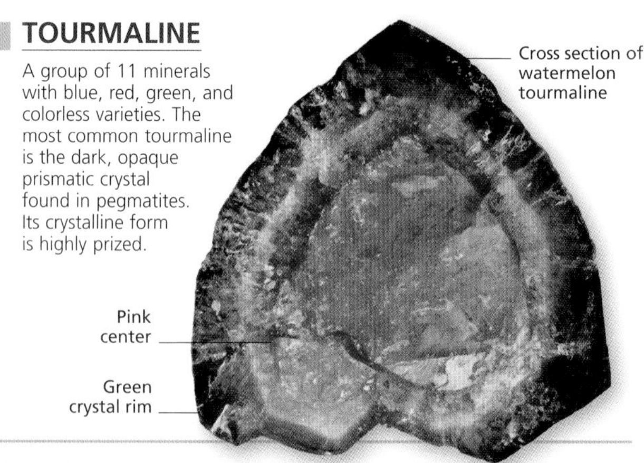

BERYL

Silicate mineral containing the metal beryllium. Pure beryl is a yellowish transparent crystal, but its impure varieties are better known. It is known as the green gem emerald when chromium is present. Its other varieties are the semiprecious blue-green aquamarine and the pink morganite.

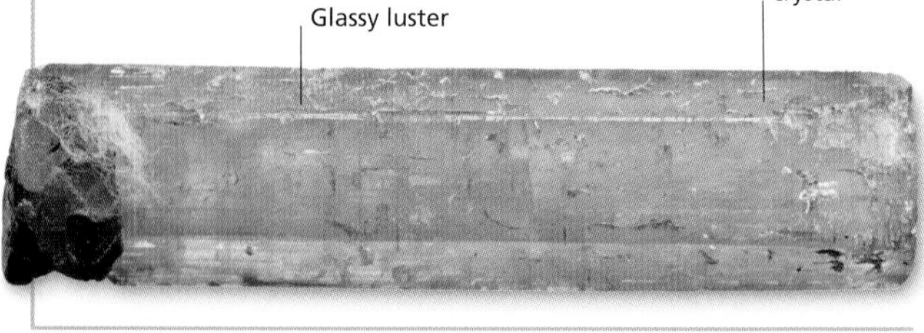

EPIDOTE

Easily recognizable by its shades of green, ranging from near black to olive green. Epidote grows as prismatic crystals, often with striations, or grooves, along their length. Specimens can also appear needlelike or as massive aggregations.

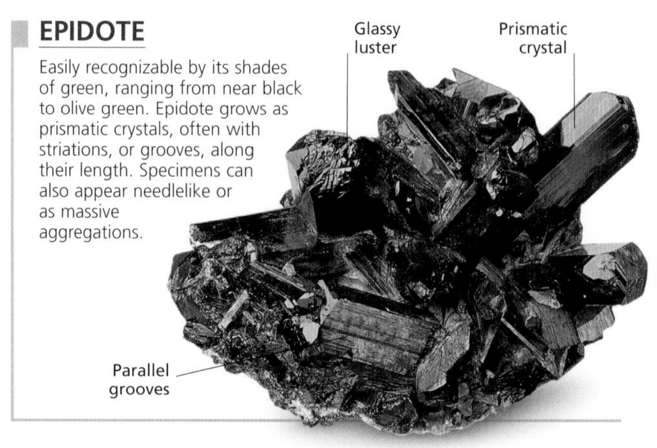

OLIVINE

One of the most abundant minerals on Earth, forming most of the rocks deep down in the upper mantle. Usually pale green, olivine crystals are wedge-shaped or tabular. Granular masses are more common than crystals. It is better known for its gem variety, peridot.

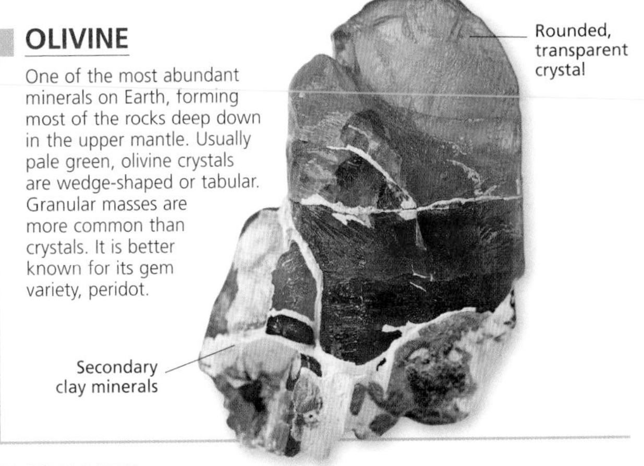

KYANITE

This attractively colored mineral gets its name from the Greek for "deep blue." It can also occur as orange, green, or colorless variants. Kyanite crystals can appear bladelike or occur in columnar or radiating forms.

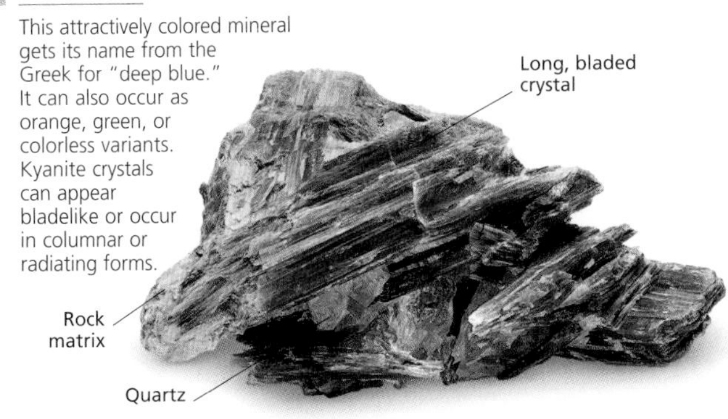

GARNET

The name given to a group of minerals, all generally a striking dark red. The most common garnet, called almandine, can be red, pink, or violet. Garnet usually occurs as well-developed crystals, often embedded in rocks.

Sulfides

These minerals contain the non-metal element sulfur, often combined with one or two metals. This makes sulfides important ores—sources of valuable metals.

GALENA

Heavy, dark gray mineral often with well-defined cubic crystals and a metallic sheen. Galena is the main ore of lead and is frequently found with calcite (p.96) and quartz (p.76). It is soft enough to be scratched by a coin.

Cubic crystal

Metallic luster

CINNABAR

Distinctive red mineral usually seen as a granular mass or powdery coating on a rock matrix, often with calcite (p.96). Cinnabar is the main ore of mercury and is highly toxic. Crystals of cinnabar are very rare.

ORPIMENT

Golden yellow mineral containing arsenic, a highly toxic element. It is typically found as powdery crusts on volcanic rocks and around hot springs. However, it can also form striking massive specimens with a shiny, resinous surface similar to gold (p.103).

CHALCOPYRITE

A metallic yellow mineral frequently found surrounded by quartz crystals (p.76). Chalcopyrite is an important ore of copper (p.102). Often mistaken for pyrite (p.92), it has pyramidal crystals and a more brassy hue.

»

PYRITE

Golden yellow, iron-rich mineral. Pyrite often occurs as well-formed cubic crystals embedded in a rock matrix. It also has an eight-sided, or octahedral, habit. When it occurs as granular nodules, it may be mistaken for nuggets of gold (p.103), which is why it is famously known as "fool's gold."

STIBNITE

Ore of the heavy metal antimony. Stibnite forms as a tangle of long, sometimes crooked, crystals, which are dark gray with a metallic luster. It frequently has a radiating habit with slender, needle-shaped crystals. It can also be massive or granular, or occur in sheets.

Oxides

Oxygen is the most common element in minerals and rocks. Oxides are minerals formed when oxygen bonds with other substances—usually metals and semimetals.

RUTILE

Titanium ore found as needle-shaped crystals. Rutile is generally red but can also be yellow, brown, or black. When embedded in quartz (p.76), it appears golden. The mineral is also found in massive and radiating habits.

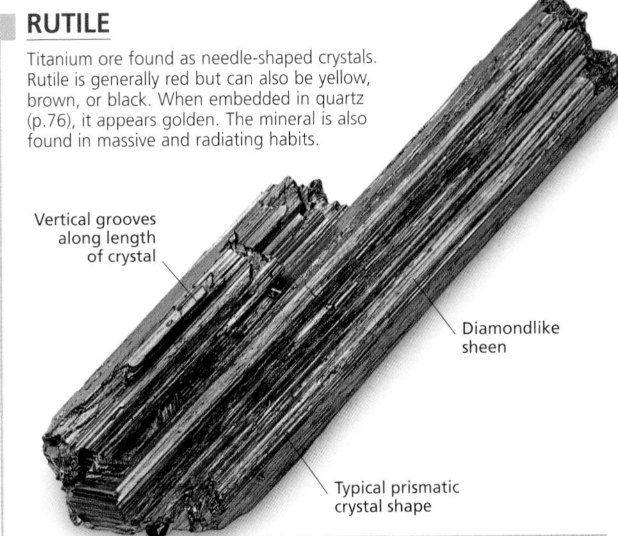

HEMATITE

Iron ore found in a number of forms, from rhombohedral crystals to kidney-shaped (botryoidal), platy, and granular masses. Crystalline hematite is dark brown or black and has a metallic luster. Botryoidal hematite is dull red.

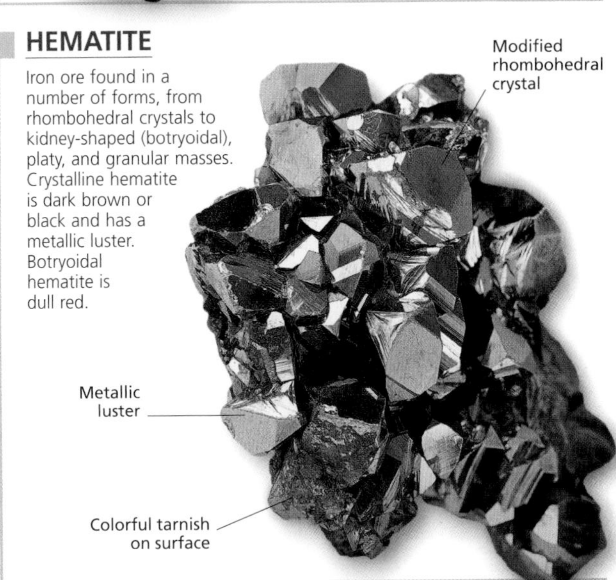

MAGNETITE

Black magnetic mineral and iron ore with a soft metallic luster. Magnetite can form octahedral (8-sided) or dodecahedral (12-sided) crystals, but is commonly seen in massive or granular forms and as crystal clusters. It also forms most black sands.

CORUNDUM

A colorless mineral better known for its gem varieties of sapphire and ruby. Second only to diamond in hardness, the red ruby variety contains chromium impurities, while the blue sapphire contains titanium and iron. Corundum crystals are hexagonal, tapering to a point, and sometimes have two pyramid-shaped ends.

SPINEL

Hard mineral with well-formed, eight-sided crystals and a long history as a gemstone. Spinel occurs in many shades, including blue and green, but it is best known in its red form, which was mistaken for ruby until the 19th century. Spinel is found as large crystals in igneous rocks. It can also exist in granular forms.

Eight-sided spinel crystal

Quartz matrix

ICE

All naturally occurring nonorganic solids are classed as minerals, making ice the most common mineral on Earth's surface. It is mostly found in massive aggregates, although snowflakes are crystalline. Pure ice is blue, but when air bubbles become trapped inside ice they give it a white, frosted color.

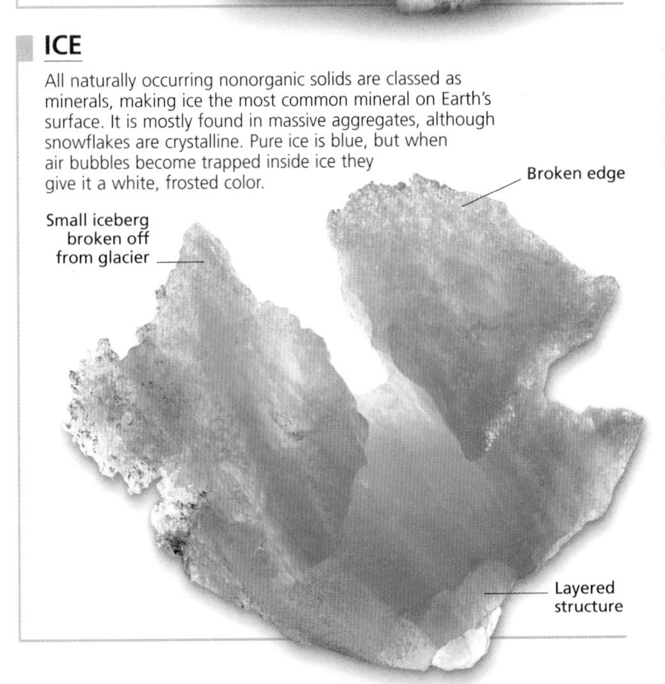

Carbonates

Carbonate minerals contain carbon and oxygen in a ratio of 1:3. These minerals react readily with some acids, are generally soft, and wear away easily.

CALCITE

The most common calcium carbonate mineral. It is white or colorless when pure, and pinkish red and brown when it contains impurities. Spiked crystal specimens are called dogtooth spar, while those with flat-topped, columnar shapes are known as nailhead spar. Calcite is found in limestones, chalks, and marbles.

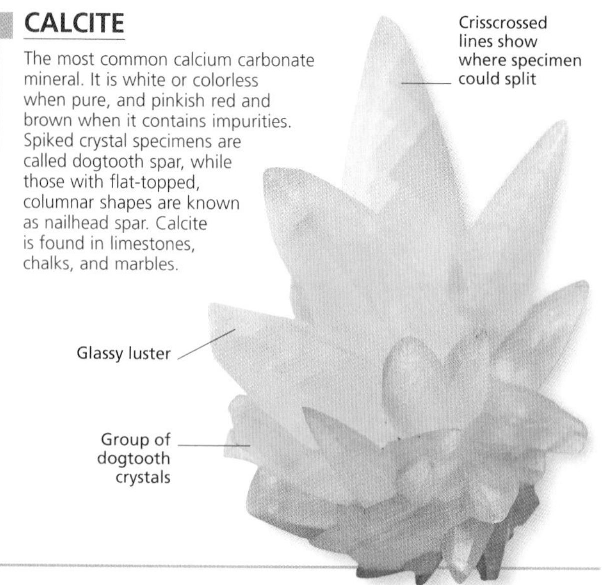

DOLOMITE

Carbonate mineral ranging from colorless and white to brown, gray, and red. Dolomite contains magnesium as well as calcium. It is generally found in darker limestones. Although typically granular or massive, dolomite can also occur as large, rhombohedral crystals.

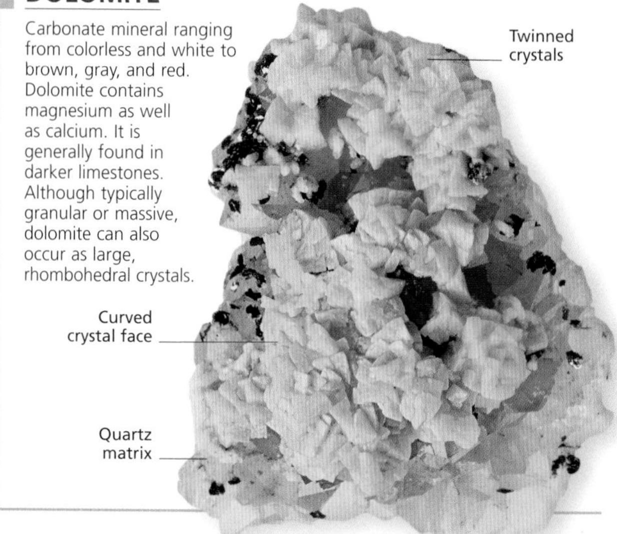

ARAGONITE

Carbonate with crystal habits ranging from tabular to needlelike, prismatic, and radiating. It is the main mineral in speleothems—rocky cave formations such as stalactites. Aragonite has the same chemical formula as calcite (opposite), but a different molecular structure. Specimens can be white, yellow, green, red, blue, or brown.

MALACHITE

Distinctive green copper ore. This mineral is found in botryoidal form—globular, like a cluster of grapes—or as a prickly, fibrous crust on a rock. Polished malachite frequently has delicate green banding. Single crystals are prismatic.

Organics

Organic substances—anything produced by or made of living things—can form rocks and are collected as gems, just like inorganic minerals.

AMBER

Soft, fossilized tree resin that fractures to form a sharp edge. Often found in areas rich in coal, amber is noncrystalline and occurs in nodules. It is generally orange, with a resinous surface.

PEARL

Ball of aragonite (p.97)—a form of calcium carbonate—that forms inside the shells of oysters and other bivalve mollusks. The shell's inner surface is coated with a mix of calcium carbonate and a small amount of conchiolin—an organic substance—to form a layer known as mother-of-pearl. This material collects around a sand grain trapped in the shell to form a pearl.

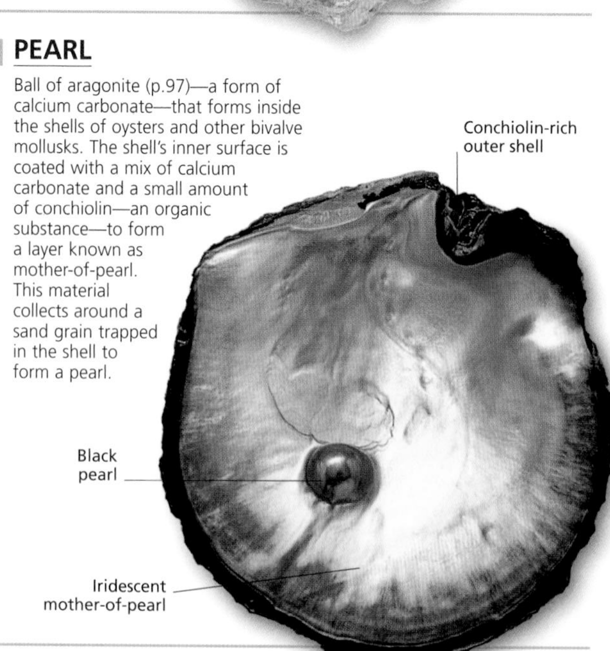

SHELL

Made from calcium carbonate, a mixture of calcite and aragonite. In mineralogical terms, shell refers to the protective armor of mollusks, such as scallops. A sediment of shells from dead mollusks, crushed up by turbulent waters, forms some limestones.

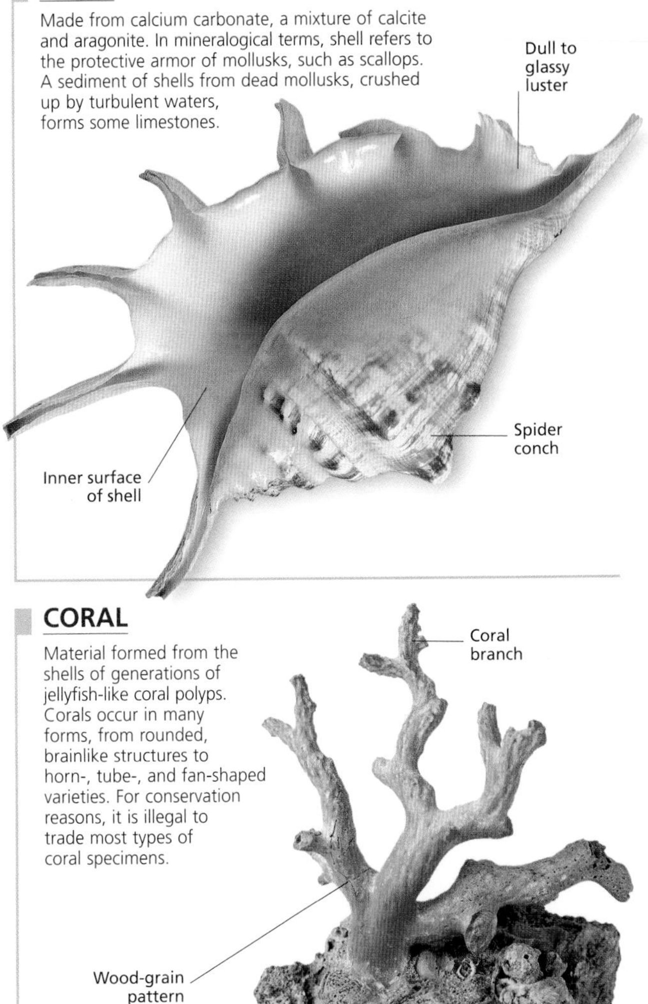

CORAL

Material formed from the shells of generations of jellyfish-like coral polyps. Corals occur in many forms, from rounded, brainlike structures to horn-, tube-, and fan-shaped varieties. For conservation reasons, it is illegal to trade most types of coral specimens.

Fossils

Millions of years old, fossils show how life in the past differed from the plants and animals that inhabit Earth today. Geologists also use fossils to date rocks.

Fossils are almost exclusively found embedded in sedimentary rocks. The most common fossils are those of aquatic plants and animals. The fossilization process (opposite) always goes through the same stages, although the time taken may vary greatly.

Fossilized turtle
This turtle fossil dates back to the time of the dinosaurs. After the creature was buried, its soft parts decomposed, while the shell and bones were replaced with new minerals and cemented into stone.

What is a fossil?

Any evidence of a living thing left in rock is a fossil. This could be the impression of a body, a footprint, an egg, a nest, or even feces. Hard body parts—bones or shells—of animals buried in rock eventually turn to stone.

How fossils form

After an organism dies, its body decays. However, if it is buried in mud, its hard parts usually remain intact while the sediment turns to stone. Sometimes fossils form when the original minerals are replaced by others carried in the water that saturates the sediment. Often, the body parts are gradually washed away, leaving a mold of the original, which may be filled with new minerals to make a cast fossil. The fossil is only revealed millions of years later, when the softer rocks around it erode.

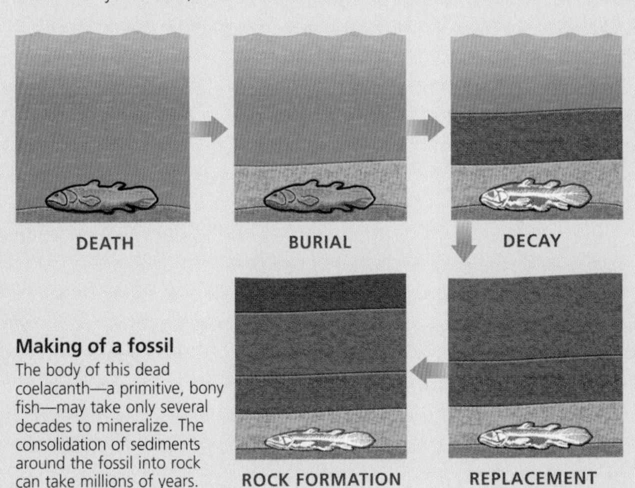

Making of a fossil
The body of this dead coelacanth—a primitive, bony fish—may take only several decades to mineralize. The consolidation of sediments around the fossil into rock can take millions of years.

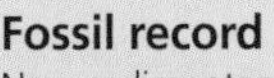

Fossil record

New sedimentary rocks form as layers (strata) on top of older rocks. So, the deeper a fossil is in the layers of sedimentary rock, the older it must be. This fact can be used to construct a fossil record that traces the evolution of life on Earth. Some fossils are found only in rocks of a certain age. Called index fossils, they are useful in determining the age of a rock.

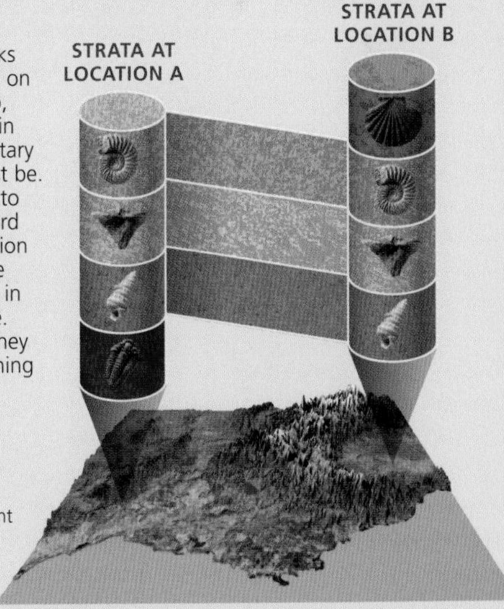

Strata dating
By comparing fossils found in rocks at different locations (as shown here), it is possible to work out the relative ages of rock layers.

Native Elements

While most elements occur in a combined state within minerals, native elements are found in a pure form. This means they cannot be divided into simpler ingredients.

SULFUR

Distinctive yellow mineral that melts to become red. Pure sulfur is found in volcanically active areas. It can form resinous crystals shaped like needles or rectangular prisms but is generally found as powdery blocks.

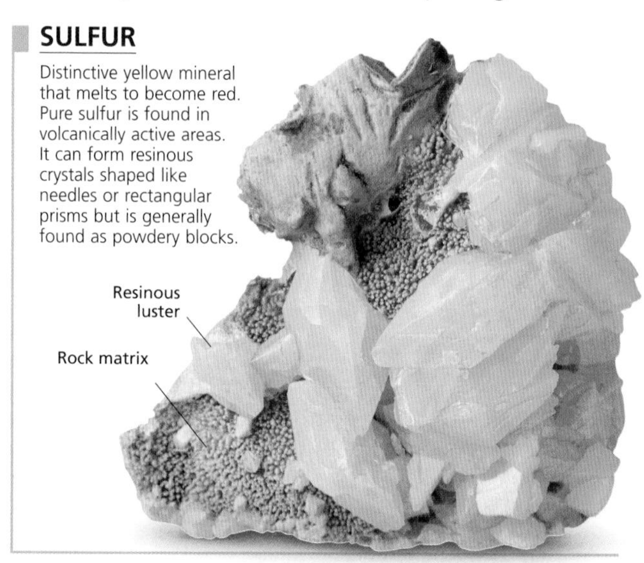

COPPER

Pinkish red metal rarely found in a pure state. Metallic copper is found in veins or mixed with iron- and copper-rich minerals. It forms flat masses or is branching, spreading out in a continually dividing form. It can also be found as cubic crystals.

GOLD

Deep yellow metal, almost always uncombined. Gold occurs as nuggets and plates, but more often as grains embedded in rock. It is nonreactive and keeps its luster even after millions of years of having been buried underground.

SILVER

Metallic white element found in rocks where hot waters once trickled. Deposits of pure silver are wiry or branching and seldom shiny. Silver's surface turns black or brown when it comes into contact with other chemicals.

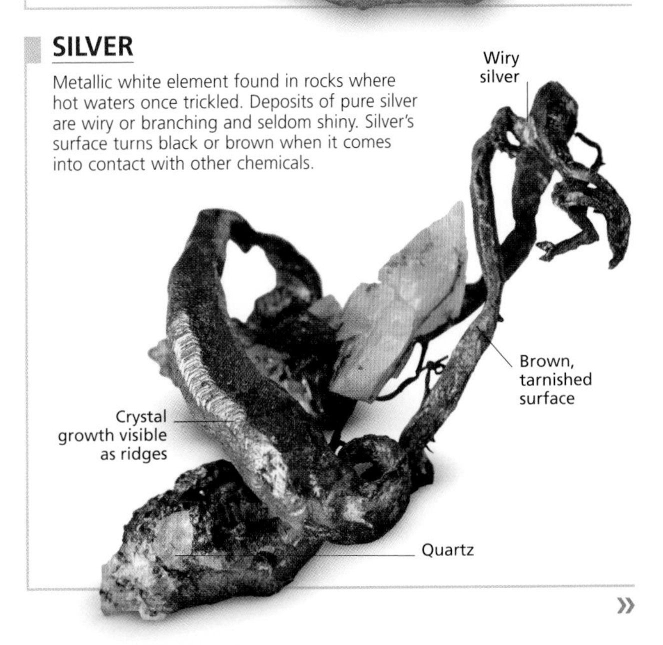

»

GRAPHITE

Metallic-looking form of pure carbon, which takes shape as shiny lumps with a slippery feel. In contrast to diamond (below), graphite is very soft and its surface scrapes off easily. It is commonly used as pencil lead.

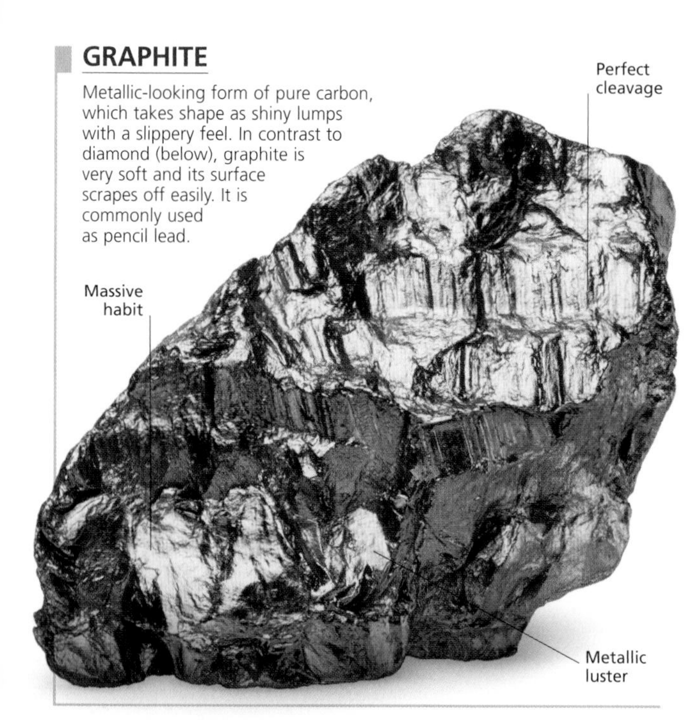

DIAMOND

Colorless form of carbon and hardest of all minerals. It is prized as a gem. Natural diamonds occur as crystals, sometimes tinged with pinks, blues, and yellows due to impurities. Flecks of carbon can often be seen in crystals. Low-quality, gray-black diamonds are used in industrial cutting tools.

Other Minerals

These include specimens from smaller and rarer mineral groups, such as halides (containing halogens), borates (containing boron), and sulfates and phosphates.

BORAX

Borate mineral associated with deserts and salt flats. Most specimens are anhydrous, meaning that their water content has evaporated, leaving a yellow-gray earthy mass. However, it is colorless when its crystals contain trapped water molecules. Rarely, crystals are tabular.

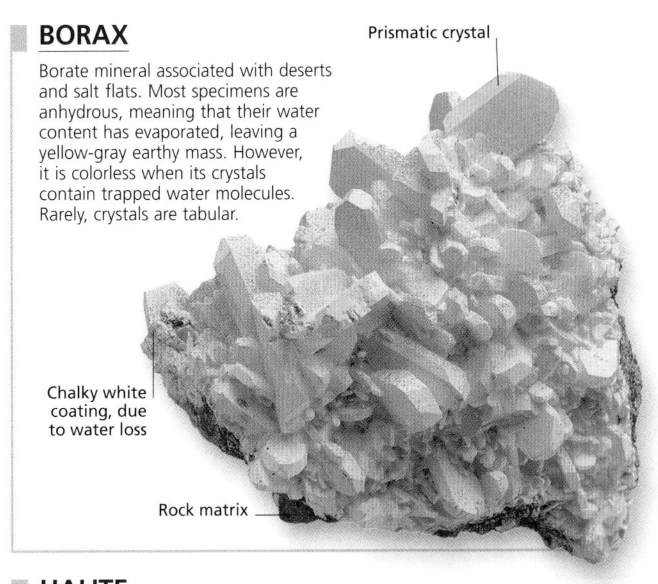

HALITE

Mineral form of table salt, generally found in massive aggregates of rock salt (p.54) or granules left by evaporated saltwater. It is colorless when pure, and reddish or orange in most rock salts. Halite forms cubic crystals and is unusual in that it dissolves in water.

»

GYPSUM

Colorless or white, with a tinge of color when impure. In its dried form it is a component of plaster of Paris and solidifies upon the addition of water. Its desert rose form can be made of slanted blades, prismatic columns, or delicate fibers.

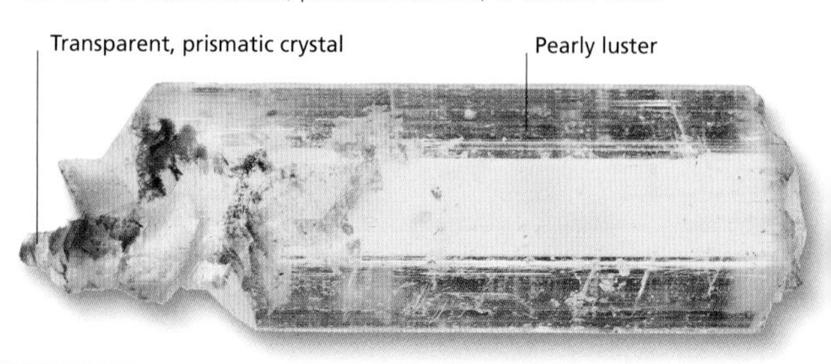

BAUXITE

A mixture of several aluminum ores that occurs as clay-rich, earthy deposits, nodules, and pisoliths—spongelike rocks filled with pea-sized grains. It is generally yellow or brown but can also be pink or red.

GOETHITE

Mineral form of rust, typically red-brown but can be darker depending on crystal size—smaller crystals are light, larger ones are dark. Goethite forms velvet crusts, radiating fibers, flat tabular crystals, and globules.

APATITE

Mineral similar to the stiffening material found in bones and teeth. Apatite forms glassy, prismatic crystals or waxy, compact masses and nodules. Its hydrated forms are yellow, while other types are brown, pink, or rosy red.

TURQUOISE

Named for its major trading center, Turkey, this microcrystalline mineral is often carved into figurines and cabochons. Turquoise is found as veins running through rocks or amorphous nodules. It has a waxy luster but can be polished to a shine.

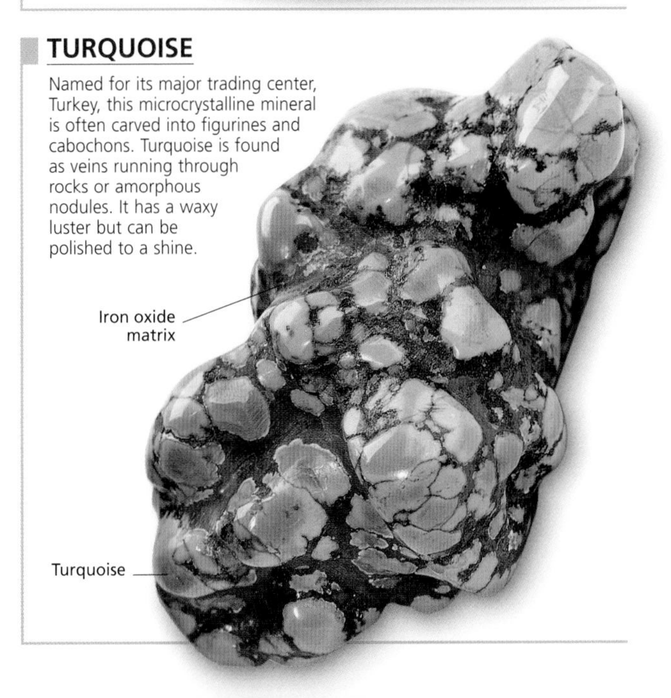

ROCK GALLERY

This gallery displays the rocks profiled in this book, grouped by type: igneous, sedimentary, or metamorphic. Within each group, the rocks are arranged by grain size, starting with coarse-grained and ending with fine-grained. You can use this gallery to see how rocks that vary in appearance, and feature in different parts of the book can belong to the same type.

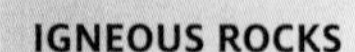

IGNEOUS ROCKS

Graphic granite
p.18

Pink granite
p.19

Granitic pegmatite
p.19

Kimberlite
p.22

Pyroxenite
p.22

Granite
p.24

Tourmaline pegmatite
p.24

Diorite
p.25

Granodiorite
p.26

Anorthosite
p.31

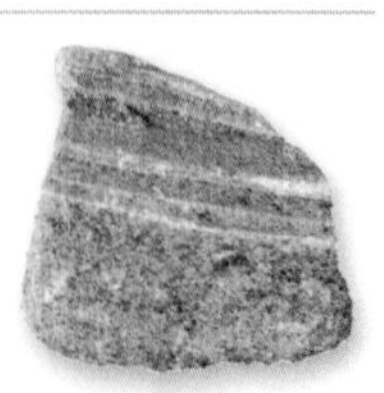

Tuff
p.31

Gabbro
p.38

Peridotite
p.38

Dolerite
p.39

Lamprophyre
p.39

Layered gabbro
p.42

Leucogabbro
p.42

Carbonatite
p.43

Porphyry
p.43

Syenite
p.46

Scoria
p.50

Pumice
p.50

Trachyte
p.51

Basalt
p.57

Vesicular basalt
p.57

Andesite
p.58

»

»

Obsidian
p.58

Rhyolite
p.59

Volcanic bomb
p.59

Amygdaloidal basalt
p.71

Dacite
p.72

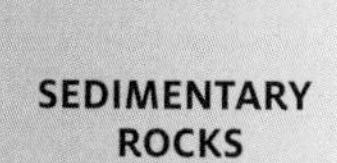

SEDIMENTARY ROCKS

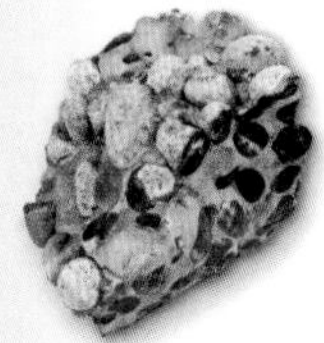

Conglomerate
p.27

Breccia
p.27

Greensand
p.33

Arkose
p.33

Gritstone
p.35

Sandstone
p.34

Red sandstone
p.34

Oolitic limestone
p.35

Orthoquartzite
p.41

Ironstone
p.41

Limestone
p.52

Tufa
p.52

Dolomite
p.53

Chalk
p.53

Green marl
p.54

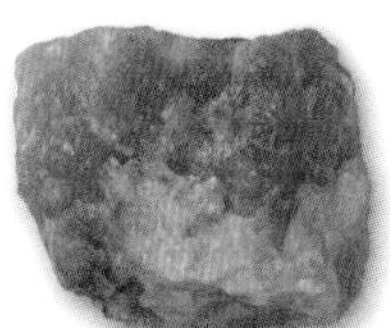
Rock salt
p.54

Rock gypsum
p.55

Diatomite
p.56

Loess
p.56

Red marl
p.64

Shale
p.64

Oil shale
p.65

Mudstone
p.66

»

Siltstone
p.67

Graywacke
p.67

Tillite
p.68

Flint
p.68

Chert
p.69

Bog iron
p.69

Coal
p.70

Laterite
p.71

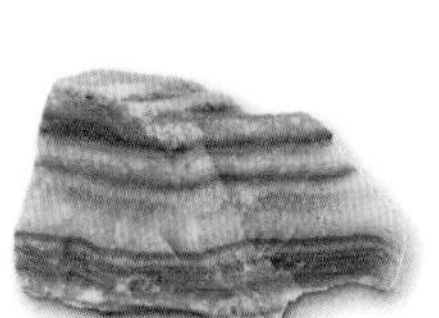
Travertine
p.72

Banded ironstone
p.73

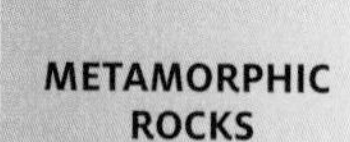

Serpentinite
p.23

Eclogite
p.23

Amphibolite
p.25

Quartzite
p.30

Marble
p.32

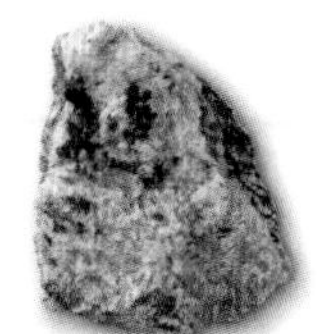

Kyanite schist
p.32

Muscovite schist
p.40

Folded schist
p.40

Migmatite
p.44

Granulite
p.45

Skarn
p.45

Banded gneiss
p.46

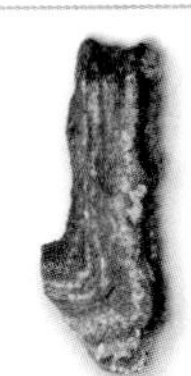

Folded gneiss
p.47

Soapstone
p.51

Hornfels
p.60

Blueschist
p.60

Slate
p.61

Phyllite
p.61

Mylonite
p.73

MINERAL GALLERY

This gallery presents the minerals profiled in this book. They are arranged according to their mineral families, which are defined by their chemical constituents. It is not possible to include all minerals—there are thousands of them, arranged in dozens of families and groups—but this section allows you to see the great variety of mineral colors and shapes.

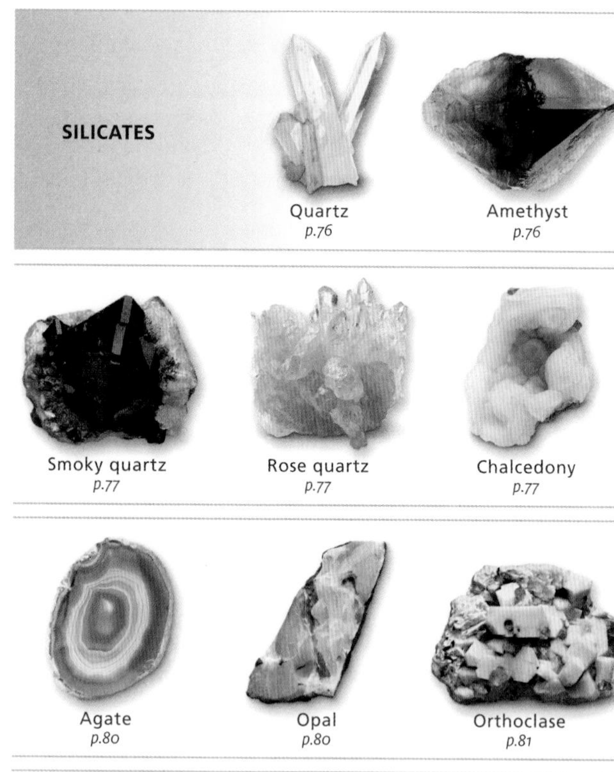

Quartz *p.76*

Amethyst *p.76*

Smoky quartz *p.77*

Rose quartz *p.77*

Chalcedony *p.77*

Agate *p.80*

Opal *p.80*

Orthoclase *p.81*

Microcline *p.81*

Anorthoclase *p.82*

Albite *p.82*

Oligoclase
p.82

Anorthite
p.83

Labradorite
p.83

Andesine
p.83

Serpentine
p.84

Chrysotile
p.84

Talc
p.84

Mica
p.85

Biotite
p.85

Bentonite
p.86

Kaolinite
p.86

Diopside
p.87

Augite
p.87

Hornblende
p.87

Tourmaline
p.88

»

»

Beryl
p.88

Epidote
p.88

Olivine
p.89

Kyanite
p.89

Garnet
p.89

SULFIDES

Galena
p.90

Cinnabar
p.90

Orpiment
p.91

Chalcopyrite
p.91

Pyrite
p.92

Stibnite
p.92

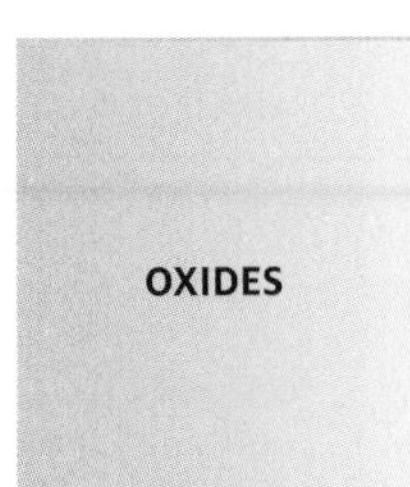

OXIDES

Rutile
p.93

Hematite
p.93

Magnetite
p.94

Corundum
p.94

Spinel
p.95

Ice
p.95

CARBONATES

Calcite
p.96

Dolomite
p.96

Aragonite
p.97

Malachite
p.97

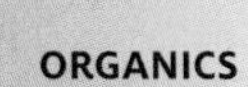

ORGANICS

Amber
p.98

Pearl
p.98

Shell
p.99

Coral
p.99

NATIVE ELEMENTS

Sulfur
p.102

Copper
p.102

Gold
p.103

Silver
p.103

Graphite
p.104

Diamond
p.104

OTHER MINERALS

Borax
p.105

Halite
p.105

Gypsum
p.106

Bauxite
p.106

Goethite
p.106

Apatite
p.107

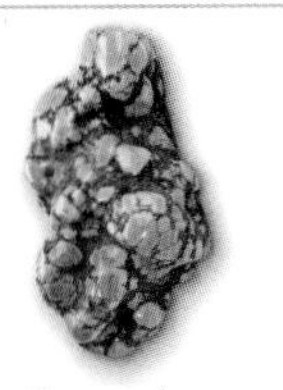

Turquoise
p.107

Glossary

The study of rocks and minerals has its own jargon. A few terms can help you understand rocks and minerals better and identify them with greater precision.

Aggregate A large cluster of similar mineral crystals and rock fragments.

Alkali feldspar A type of silicate mineral that contains a high percentage of potassium and sodium.

Bed The smallest division of a sedimentary rock formation. Each bed indicates a change in the original material being deposited.

Bedding plane The surface that separates adjacent beds. All beds are originally horizontal but are often rotated, twisted, or buckled over the years.

Bladed A crystal habit in which the crystals resemble a blade.

Botryoidal A habit in which the mineral occurs in globular masses—and resembles a bunch of grapes.

Breccia A sedimentary rock consisting of angular fragments of older rocks.

Cabochon A gem that has been cut and polished to give it a domed top and a flat bottom.

Carbonate A compound that contains carbon and oxygen atoms in a 1:3 proportion.

Clast A fragment of rock or mineral that becomes part of a new sedimentary rock.

Clay A member of a large group of silicate minerals that occur as very fine grains.

Cleavage The way a mineral breaks into sections with flat surfaces. Some minerals can do this, others cannot.

Columnar A habit in which the mineral forms into thick, elongated columns.

Conchoidal fracture A property of certain rocks and minerals that have a curved or shell-like fracture. Flints break in this way.

Cross-bedding Layering in which neighboring beds are orientated in different directions—one may be horizontal, the other slanted at an angle.

Crystal A mineral in which the inner symmetry of its atomic arrangement is reflected in flat outer faces in a geometric pattern.

Cut The particular shape and geometric form given to a gem.

Dodecahedral A term used to describe a crystal having 12 sides.

Dull luster A type of luster in which little light is reflected.

Earthy luster A type of luster in which the mineral's surface looks powdery or granular.

Erosion A process in which fragments of rock and mineral are loosened and carried away by wind, water, or ice.

Eruption A natural phenomenon in which volcanic material—such as magma, ash, and gas—bursts out of Earth's surface through a volcano or vent.

Extrusive rock An igneous rock that forms on the surface of Earth.

Face A flat surface of a crystal.

Feldspar A group of common silicate minerals that are found in many intrusive igneous rocks. Feldspar means "field crystal" in German.

Fissile texture A rock texture which allows the rock to be split into sheets.

Flow banding Marks in an igneous rock that show that the molten lava or magma was flowing until it cooled and became a solid rock.

Foliation A characteristic of metamorphic rocks whereby the component crystals have been arranged into sheets or waves.

Fossil Evidence of life left in a rock.

Fracture The way that rocks and minerals break.

Garnet A member of a group of silicate minerals that often contain magnesium, calcium, and iron.

Gem/gemstone A precious crystal or stone used in decoration, especially in jewelry.

Gem cutting The process of shaping a gem to maximize its sparkle and color.

Glass A noncrystalline solid in which the molecules are jumbled in a random order, rather than arranged in a repeating pattern, as in crystals.

Glassy texture The smooth and shiny surface appearance of a rock or mineral, similar to that of a piece of glass.

Grain The basic unit that makes up a rock.

Granular texture A mineral texture in which the grains are visible to the naked eye.

Groundmass see *matrix*

Habit The large-scale appearance of a mineral. Some minerals occur in a range of habits.

Hackly fracture The fracture of a mineral, often a metal, that breaks into jagged sections.

Heavy metal A metal that is usually dense and heavy. Examples include mercury and antimony.

Igneous rock A rock that forms when melted rock called magma cools down and solidifies.

Inclusion A crystal within another crystal. The two crystals are usually of different minerals.

Intrusive rock An igneous rock that cools from magma under the surface of Earth.

Iridescence The reflection of light from internal elements of a stone, yielding a rainbowlike play of colors.

Lava Magma that has reached the surface.

Luster A characteristic of a mineral describing how light interacts with the surface, giving it a particular appearance.

Magma Molten or liquid rock that forms deep beneath Earth's surface, where it is hot enough to melt minerals.

Massive A mineral habit in which the material has no definite form and individual crystals cannot be seen.

Matrix The fine-grained background material of a rock into which or on top of which larger crystals are set.

Metal A substance characterized by high electrical and thermal conductivity as well as by malleability, ductility, and high reflectivity of light.

Metallic luster The shiny, reflective surface appearance of certain minerals, resembling that of a metal.

Metamorphic rock Rock that forms when another rock is subjected to heat and pressure underground, but without remelting.

Mica A member of a series of silicate minerals, frequently occurring in thin sheets, which are common components of rocks.

Microcrystalline A mineral with crystal grains too small to see with the naked eye.

Mineral group A family of minerals that share a certain chemical composition or structure.

Native element A naturally occurring pure chemical element that cannot be separated into simpler components. Examples include gold and sulfur.

Nodule A generally rounded accumulation of sedimentary material differing from its enclosing sedimentary rock.

Nonmetal A material that does not have metallic properties.

Octahedral Having eight sides arranged as two base-to-base pyramids.

Ooliths Individual, spherical sedimentary grains, usually of calcite, from which oolitic rocks are formed.

Ore A mineral that is a valuable source of useful materials, such as metals.

Oxidation The process by which a substance reacts with oxygen in air and water.

Pegmatite An intrusive igneous rock in the form of a vein, and composed of large crystals.

Phenocryst A large crystal set into an igneous rock composed of a matrix of smaller grains.

Platy A mineral habit in which the crystals occur in plates.

Porphyritic texture The texture of a rock studded with large crystals.

Porphyroblast A large single crystal that forms within a metamorphic rock.

Precious A term referring to the most highly prized minerals and metals.

Prismatic Having the shape of a prism.

Pyramidal Having the shape of a pyramid.

Pyroclastic rock A rock formed from the ash and other material forced out of a volcano in an explosive eruption.

Pyroxene Part of a group of 21 important rock-forming silicate minerals containing silicon, oxygen, and metals such as calcium, aluminum, and magnesium among many others.

Radiating habit A type of habit in which the crystals have grown outward from a common center in radiating form.

Refractive index A figure that shows how much a beam of light is bent as it shines through a particular transparent solid.

Resinous luster The shine of resin—a soft sheen combined with a greasy appearance.

Rhombohedral Having the shape of a rhombus, or parallelogram.

Rock dating The process of determining the age of a rock.

Semimetal A material that exhibits characteristics of both a metal and a nonmetal. Silicon is the most common semimetal.

Semiprecious A term describing a gem that is highly prized but less valuable than the so-called precious stones.

Sedimentary rock A rock formed on Earth's surface from layers of sediment deposited on each other.

Silicate A mineral compound containing silicon and oxygen and usually a number of other elements. Most rocks are made of silicate minerals.

Stalactite A rock formation that hangs down from the roof of a cave.

Stalagmite A rock formation rising up from the floor of a cave.

Striations Parallel horizontal or vertical grooves or lines appearing on a crystal.

Tabular A mineral habit in which the crystals occur in flat blocks.

Tarnish The process in which a rock's metallic luster is diminished by the oxidation of the surface, which reduces its shine and reflectivity.

Termination Faces that make up the ends of the crystal.

Texture A description of the look and feel of the surface of a rock.

Trace element An element that occurs in very small amounts in a mineral.

Twinned crystals Two crystals of the same mineral that share a face and grow away from each other in opposite directions.

Variety A form of a mineral with a different appearance but the same chemical composition.

Vein A geological feature where a thin, sheetlike mineral or rock fills a crack running through another rock.

Vesicle A spherical or oval cavity produced by a gas bubble in magma that remains after the magma has solidified to form a rock.

Weathering A process by which wind, water, ice, natural acids, or the action of animals and plants break rocks into fragments.

Index

Page numbers in **bold** indicate main entry.

A

B

C

D

E

F

G

P

Q

R

S

T

V W